The Eternal City

The Eternal City

A History of Rome in Maps

Jessica Maier

The University of Chicago Press   *Chicago and London*

The University of Chicago Press, Chicago 60637
The University of Chicago Press, Ltd., London
© 2020 by The University of Chicago
All rights reserved. No part of this book may be used or repro-
duced in any manner whatsoever without written permission, ex-
cept in the case of brief quotations in critical articles and reviews.
For more information, contact the University of Chicago Press,
1427 E. 60th St., Chicago, IL 60637.
Published 2020
Printed in China

29 28 27 26 25 24 23 22 21 20   1 2 3 4 5

ISBN-13: 978-0-226-59145-2 (cloth)
ISBN-13: 978-0-226-59159-9 (e-book)
DOI: https://doi.org/10.7208/chicago/9780226591599.001.0001

Library of Congress Cataloging-in-Publication Data

Names: Maier, Jessica, author.
Title: The Eternal City : A History of Rome
in Maps / Jessica Maier.
Description: Chicago : The University of Chicago
Press, 2020. | Includes bibliographical references and index.
Identifiers: LCCN 2020003837 | ISBN 9780226591452
(cloth) | ISBN 9780226591599 (ebook)
Subjects: LCSH: Cartography—Italy—Rome. | Rome (Italy)
—Geography. | Rome (Italy)—Maps.
Classification: LCC DG806.2.M345 2020 | DDC 945.6/32—dc23
LC record available at https://lccn.loc.gov/2020003837

Contents

Contents

Contents

Introduction

Rome as Idea and Reality

"Now let us, by a flight of imagination, suppose that Rome is not a human
habitation but a psychical entity with a similarly long and copious past—
an entity, that is to say, in which nothing that has once come into
existence will have passed away and all the earlier phases of development
continue to exist alongside the latest one."

Sigmund Freud, *Civilization and Its Discontents*, 1930

Visitors to Rome today usually come to see the city's ancient and Renaissance landmarks—places like the Forum, the Colosseum, St. Peter's Basilica, and the Sistine Chapel. But more than its individual sites, what makes
Rome as a whole so captivating is its unbroken history—the countless incar-
nations and eras that merge in the cityscape. No other place can quite match
Rome's resilience, its three-millennium-long series of reinventions.

Anyone looking for the quintessential Roman experience in a single spot
would do well to bypass showstoppers like the Pantheon or Trevi Fountain
and head instead to San Clemente, a relatively modest church at the foot of the
Lateran Hill in the southeastern part of the city. The building that people enter
at ground level dates to the twelfth century and is a fine example of medieval
Christian architecture (fig. 1). That church, however, is just the tip of the ice-
berg. From a door inside the gift shop you can descend through archaeological layers, down a staircase to an earlier church from the fourth century, and

Fig. 1

Basilica of San Clemente,
Rome, dedicated 1108.
Photo: Erich Lessing / Art
Resource, NY.

beneath that to a second-century pagan temple, as well as previous structures dating to the first century and possibly earlier (fig. 2). The oldest part of the complex is some two thousand years old and sixty feet beneath street level.

When you climb back up to emerge into the hustle and bustle of modern Rome—the trams, taxis, and pedestrian traffic in that busy stretch of the city— you have come full circle, in a disorienting kind of time travel. San Clemente embodies the vertical, chronological layering that is so characteristic of Rome: not one city, but many superimposed and still visible today.

At the same time, Rome is more than brick and mortar. It also exists in the realm of ideas: of history, myth, and symbolism. All these factors are shaped, in turn, by human ones. From gods to caesars, pagans to Christians, popes to prime ministers, and pilgrims to tourists, Rome has been many things to many people. This book considers the city through the eyes of artists and mapmakers who have managed to capture something of its essence over the centuries. For all of them, the key question has been which slice of Rome to show. The dilemma is summed up by the almost hallucinogenic cover to a late twentieth-century book on the city's development (fig. 3), which squeezes 2,700 years of history into a two-dimensional diagram. Aiming to be comprehensive, it becomes overwhelming, like a restaurant menu with too many choices. Or perhaps it is deliberately disorienting, in a tacit acknowledgment of the futility of condensing the multiplicity that is Rome into a single image.

The maps in this book, by contrast, provide the raw material for a selective visual history of not one but ten "Romes," each reflecting a key theme or era. The aim is not to provide an encyclopedic overview of all maps of Rome throughout history, but rather to focus on a judiciously chosen selection that best illuminates key facets of the Eternal City. The chapters are arranged chronologically, while the maps are drawn from across time. Each chapter begins with a historical introduction to set the stage, describing the relevant cultural background that helps us to better understand the maps, which in turn help us to better understand Rome itself. Descriptions and interpretations of the individual maps follow the chapter introductions, each one investigated for all the insights it has to offer.

Collectively, the maps chosen for inclusion in this book introduce us to different cities that flourished on the same site at different times, but always overlapped with or built on previous versions—as, for example, when Fascist-era urban planners looked for inspiration to the ancient city, much as their Renaissance counterparts had done four hundred years earlier. Or, for that matter, when the medieval builders of San Clemente chose to construct a new church on old foundations. Rome, it seems, is always referring back to itself.

The city is and has long been a shape-shifter, its very adaptability allowing it to live on and remain relevant for so long. But to a significant extent Rome's previous lives are always present, never fully wiped away. The images in this book effectively convey that rich existence. Ranging from modest to magnificent, they comprise singular aesthetic monuments like paintings and grand prints as well as more popular and practical items like tourist maps, geological and archaeological surveys, and digitizations. The most iconic, important, and beautiful images of the city appear alongside relatively obscure, unassuming items that have just as much to teach us about the Eternal City.

Legend holds that Rome was founded in 753 BCE by Romulus on the Palatine Hill, the first of the city's famed seven hills—a place where excavations have indeed yielded evidence of modest inhabitation around that time (fig. 4). It is likely that separate settlements popped up on other nearby hilltops around the same time. Yet Rome's advantageous site probably attracted settlement even earlier. Located at a crossroads and in a valley, it was a fertile spot for agriculture, the river and hills provided natural defenses, and the island in the Tiber was a strategic lookout point for monitoring river traffic. That said, the place also had pronounced disadvantages. Prone to flooding, much of it was a marshy plain. Little in Rome's site predicted the great city it was to become.

According to ancient authors, a series of kings ruled the Roman people, who merged over time from several ethnic groups—including Sabines, Latins, and

Fig. 2

Temple of Mithras in the lower church of San Clemente, Rome, ca. 200 CE. Photo: Erich Lessing / Art Resource, NY.

Etruscans. The city was a republic from about 500 BCE until the first century BCE, during which time its architecture and infrastructure developed rapidly, as did its power over the surrounding region of Latium, the larger Italian peninsula, and eventually the Mediterranean world. By the time the city became capital of an empire in the first century BCE, it had a population estimated at one to two million inhabitants. Imperial Rome flourished until Constantine left to found a new capital at Constantinople in the early 300s, leaving a power vacuum and sending the city into a tailspin of sacks and foreign invasions lasting several centuries. Rome had become a deeply Christian place, however, with sacred shrines to rival its decaying pagan ones, and the popes—the city's bishops—eventually assumed temporal leadership.

The Middle Ages was a millennium of ups and downs caused by internal power struggles as well as external forces, but on the whole the medieval papacy parlayed the city's prestige as a pilgrimage destination into a new kind of power and influence. The nadir of Rome's fortunes came when the papal court departed for Avignon in the early 1300s, the city's population dropping to less than 20,000. But this moment, too, proved fleeting. With the return of the

Fig. 3

Cover, Piero Maria Lugli, *Urbanistica di Roma: Trenta planimetrie per trenta secoli di storia.* Rome: Bardi Editore, 1998.

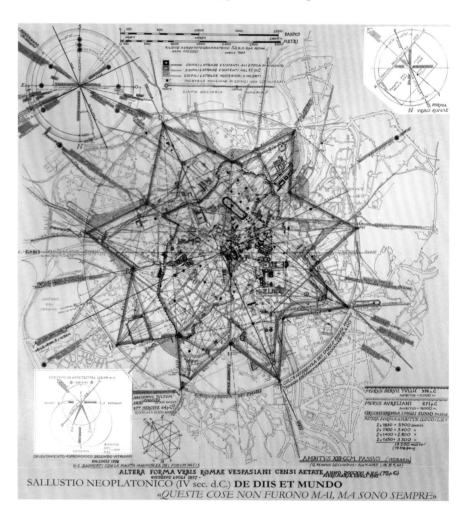

Introduction

popes in the 1400s, the city began a lasting resurgence that carried it through
centuries of political and religious challenges—the Sack of Rome by imperial
troops in 1527, the Protestant Reformation followed by the Church's Counter-
Reformation, the propagandistic glory of the Baroque period, and a new urban
identity as a tourist destination.

A fresh chapter in Roman history began in 1871, when the city was made
capital of united Italy (a time known as *Roma capitale*), and it has since been a
Fascist city, a fashion capital, a world heritage site, the home of a dysfunctional
government (or a series of them), even a place where avant-garde architecture
and design have a home.

One particularly emblematic spot is the modern building designed by Richard Meier to house the Augustan Ara Pacis, or Altar of Peace, constructed 13–9
BCE (fig. 5). Meier's museum, completed in 2006—just over two thousand
years after the structure it contains—consists of boxy expanses of concrete,
steel, glass, and travertine, the geometry of which echoes the underlying block
form of the marble altar itself. However controversial, Meier's building, together with the Augustan monument it houses, is just as much an allegory of Rome
as San Clemente. This is a city where change has been the norm for over 2,500
years, where new has responded to old, and where decline has always gone hand
in hand with regeneration.

———

An underlying premise of this book is that by understanding Rome—or trying
to—we gain some insight into cities and their patterns in general. It is a cliché
to say that we are living in an exceptionally urbanized time. More and more
people are concentrated in megacities with populations exceeding ten million—
Tokyo, Shanghai, Beijing, Karachi, Delhi, Nairobi, Lagos, Istanbul, New York,
Mexico City, and São Paulo are just some of the places that make the list. Rome
will never be as populous as those places, but it is in certain respects the mother
of all cities. At various times and for a variety of reasons, other places (Cairo,

Prague, Mecca) have laid claim to that title, but none has led such a long and uninterrupted existence, clinging to its influence as tenaciously as Rome.

Along the way, Rome has been defined—paradoxically, perhaps—by its ability to change and adapt to new circumstances. However much it has struggled to enter the modern age with its much vaunted eternity intact, Rome has important lessons to teach us about how cities become, and remain, relevant in the extreme long term. Thanks to its great staying power, Rome is also a model for how we conceptualize urban centers in general—as places defined by natural and manmade elements, but most of all by the human communities that create, inhabit, shape, and reshape them. And Rome's communities have been particularly varied. From the city's earliest days its permanent population has been joined by a shifting group of visitors, be they merchants, pilgrims, clerics, occupying armies, tourists, or diplomats.

The second basic premise of this book has to do with the nature of the material—namely, the maps and views that tell Rome's story. Such works do not necessarily aspire to factual and objective reporting. In our current era of "alternative facts" and "fake news," maps have a lot to teach us. They are necessarily selective, for they must leave out more information than they could ever possibly incorporate. Which features, streets, neighborhoods, landmarks are included—which ones omitted, as if they did not exist? The question of what makes the cut can be highly loaded, revealing tacit approval or censure, moral codes, or social critique. Maps can be actively motivated by an agenda—a way of staking a claim at the expense of competing claims. They can make

Introduction

arguments, advance falsehoods, or spread propaganda. In sum, they have the power to embody ideals and beliefs as well as (or in the face of) concrete reality.

The authoritative multivolume *History of Cartography*, a project in progress since 1987, hints at the wide range of possibilities in its encompassing definition of maps as "graphic representations that facilitate a spatial understanding of things, concepts, conditions, processes, or events in the human world." The spatial component is key, of course, but nothing else is to be taken for granted: not truthfulness, not a focus on physical features, not neutrality. For a place like Rome, where intangible factors like symbolism and memory hold as much weight as physical features, mapping might be the best way to do justice to the city—a strategy for defining the identity of a place that defies definition. Ultimately, that notion lies at the heart of the following ten chapters.

FURTHER READING

Bevilacqua, Mario, and Marcello Fagiolo, ed. *Piante di Roma dal Rinascimento ai Catasti*. Rome: Artemide, 2012.

Bogen, Steffen, and Felix Thürlemann. *Rom: Eine Stadt in Karten von der Antike bis heute*. Darmstadt: WBG, 2009.

Frutaz, Amato Pietro. *Le piante di Roma*. 3 vols. Rome: Istituto di studi romani, 1962.

Gori Sassoli, Mario, ed. *Roma veduta: Disegni e stampe panoramiche dal XV al XIX secolo*. Rome: Artemide, 2000.

Harley, J. B. "Deconstructing the Map." *Cartographica* 26, no. 2 (1989): 1–20.

Harley, J. B. "Silences and Secrecy: The Hidden Agenda of Cartography in Early Modern Europe." *Imago Mundi* 40 (1988): 57–76.

Harley, J. B., David Woodward, Matthew Edney, et al., ed. *The History of Cartography*. 6 vols. Chicago: University of Chicago Press, 1987–.

Maier, Jessica. *Rome Measured and Imagined: Early Modern Maps of the Eternal City*. Chicago: University of Chicago Press, 2015.

Marigliani, Clemente, ed. *Le piante di Roma delle collezioni private*. Rome: Provincia di Roma, 2007.

Taylor, Rabun, and Katherine W. Rinne. *Rome: An Urban History from Antiquity to the Present*. New York: Cambridge University Press, 2016.

Chapter One

———

Rome Takes Shape

"After starting from small and humble beginnings, [Rome] has grown
to such dimensions that it begins to be overburdened by its greatness."

Livy, *History of Rome*, first century BCE

Today's visitors to Rome pass through but rarely take much note of one of
the city's most impressive and well-preserved ancient monuments: the
third-century Aurelian Wall. But would Rome even be Rome without its wall?
That question raises a more basic one: What defines a city? Its skyline, its landmarks? Its history, or myths? The character of its inhabitants? The city's topography, its natural or manmade contours? Our modern notion of urban identity
is based on a number of factors, tangible and intangible, but imposed boundary
lines do not usually factor high on the list. Barring obvious border markings—
such as a river crossing, or a street sign indicating that one is "Entering X" or
"Leaving Y"—most urbanites today would struggle to pinpoint the moment
when they pass into or out of city limits. Think Las Vegas, Phoenix, or Houston:
amorphous cities defined by sprawl, not crisp edges.

But such fuzzy boundaries are a recent phenomenon. Until just a couple of
centuries ago, many cities owed their shape and much of their identity to defensive walls. These ostensibly utilitarian structures were often the most meaningful and prominent urban symbols: nothing less than the threshold between

inside and outside, self and other. Over the last couple of centuries, their role has faded. Many historical European cities traded their walled fortifications for ring roads as the era of siege warfare gave way to the era of the automobile—Vienna being perhaps the most famous example. Rome, by contrast, has kept its crisp edges intact for the better part of 1,800 years.

This chapter looks at maps from various eras that shed light on the real and symbolic importance of Rome's walls and that simultaneously allow us to trace the city's early history and growth. The first, done by geologist Giovanni Battista Brocchi in the late nineteenth century, introduces us to Rome's lasting natural features and prehistory: in other words, what was present before there was a *there* there. Paulus Merula's sixteenth-century map, in turn, encapsulates Rome's unfolding development, showing the city's earliest walled enclosure as a hilltop village, its subsequent physical expansion as a republican capital, and its continued evolution as an imperial seat, culminating in the building of the Aurelian Wall, which still defines the core urban form today. The final map, Heinrich Kiepert's nineteenth-century historical reconstruction of Rome's sequence of walled forms, illustrates the administrative districting that further carved up those physical barriers into bureaucratic units (a harbinger of the notorious administrative red tape that still characterizes life in Rome and in Italy generally).

The feature with the most lasting significance, however, is Rome's Aurelian Wall, which has remained a benchmark of the city's rising or falling fortunes for almost two millennia. Much more than just a physical attribute, it is key to Rome's history and image. Even Rome's legendary founding hinges on the laying out of walls. As recounted by a number of ancient writers, including Livy,

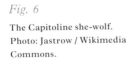

Fig. 6

The Capitoline she-wolf. Photo: Jastrow / Wikimedia Commons.

Chapter One

ORTO EX AVGVRVS DISSIDIO FRATRIBVS AD ARMA VERSSIS REMVS CÆDITVR

Fig. 7

Giovanni Battista Fontana, *Romulus Kills Remus*, plate 8, 1575. Harvard Art Museums/Fogg Museum, Anonymous Fund for the Acquisition of Prints Older than 150 Years. Photo: © President and Fellows of Harvard College.

Plutarch, and Varro, the story goes something like this: Rome was established after the fatal conclusion of a feud between Romulus and Remus, twin sons of the god Mars and descendants of Aeneas, the prince of Troy who had fled to ancient Latium following the burning of his city. This duo was a bit like a pagan Cain and Abel. As babies, Romulus and Remus had been abandoned to their fate near the banks of the Tiber. There, a fierce she-wolf discovered, suckled, and protected them until they were taken in by a shepherd and his wife, who raised them as their own (fig. 6).

As young men, the twins decided to establish a new city but argued over where to locate it—Romulus favoring the Palatine, Remus the Aventine. They resolved to settle the matter through augury, a form of prophecy relying on the observation of birds in flight, but quarreled over how to interpret the results. According to Livy's version of the story, tensions mounted until Remus, taunting his brother, leapt over the provisional wall that Romulus had marked out on the Palatine to define his new city. In revenge for that affront, Romulus killed his brother, named the city in his own honor, and made himself its king (fig. 7).

Tradition dates this fateful sequence of events to a single day: April 21, 753 BCE. Whether there is a grain of historical fact at the heart of this fable is

Fig. 8

The Servian Wall in
Termini train station,
Rome. Photo: Jeff Bondono,
www.JeffBondono.com.

the subject of some debate, but there is no denying the crucial role of walls in
Rome's mythical origins. When Romulus outlined the path of his enclosure, he
created a sacred precinct: one that no man could violate without consequence.
The physical form that Romulus was said to have given his new city was rectangular, dubbed by later generations *Roma Quadrata*. Even if no undisputed
material evidence of this legendary perimeter has come to light, it has been
equated with Rome's earliest identity as an independent city.

Rome's subsequent history can also be tracked with reference to its walls.
Livy and others relate that Rome was ruled by kings until 509 BCE, when the
citizens rose up to overthrow their rulers and institute a republican form of
government. During its first few centuries, the city fused with other hilltop
settlements and consolidated its power over the surrounding region of Latium,
expanding in terms of population, surface area, and infrastructure. As Rome
outgrew its epicenter on the Palatine, it also outgrew its original wall.

Rome's second, larger defensive circuit, the so-called Servian Wall, dates
to the fourth century BCE, although its name and possibly its original outline
derive from the time of King Servius Tullius in the 500s BCE. According to
Livy, the Servian Wall was built in the wake of Rome's brutal sack by the Gauls
in 390 BCE—an event that made it all too clear that the city was vulnerable to
foreign incursion. Constructed out of local *tufo* (or tuff, a strong but lightweight
volcanic stone, easy to quarry and widely used in Roman building projects),
its circuit spanned almost seven miles, enfolding all of the seven hills. Unlike
Romulus's wall, which has left scant if any traces in the modern city, we are on
solid archaeological footing with the Servian Wall, impressive tracts of which
still exist in Rome's eastern and southern zones. Remarkably, one section can
be seen inside the McDonald's at Termini train station, in a surreal fusion of
ancient stone and modern fast food (fig. 8).

Another half millennium would elapse before the construction of Rome's
third and final city wall. Over that time, the city changed a great deal. Already

by the third century BCE, Rome was capital of a republic that dominated the Italian peninsula and was expanding its territorial holdings throughout the Mediterranean. When Julius Caesar initiated the empire two hundred years later, Rome was the largest city in the world, highly organized and functional, with a population of more than one million.

Augustus, who ruled from 27 BCE until his death in 14 CE, famously claimed he had found Rome a city of brick and left it a city of marble. During his long, stable rule, the city gained a glittering monumentality to match its international stature. Traces of that golden age still exist in monuments like the Ara Pacis and Mausoleum of Augustus, as well as in many of the obelisks that dot the city, which the emperor looted from Egypt to adorn his capital. Conspicuously absent among Augustus's many urban embellishments is a new and improved wall. The Servian circuit was becoming a relic of a previous age, having long ago lost its usefulness as a nominal boundary. Since its construction, the city had overrun both sides of the Tiber, and its population had increased more than twenty-five-fold. Despite this rampant growth, the Servian Wall was not replaced until the waning years of the empire in the late third century.

Why? In short, a city that needed walls was vulnerable. For centuries, Rome's security had been ensured not by any tangible defensive structure but rather by the vast, buffering limits of its own empire. Insulated by its territorial holdings and watched over by its formidable army, Rome was stable, thriving, and impervious. Who needed walls? Only when infighting and external threats seemed poised to end the longstanding Pax Romana (or Roman peace) did new defenses become necessary.

Built by Emperor Aurelian in the 270s and significantly raised by Honorius in the early 400s, Rome's third and final wall is a formidable chain of brick-faced

concrete curtain punctuated by city gates and watch towers, which runs an eleven-mile circuit and reaches a height of some fifty feet (fig. 9). Paradoxically, Aurelian's wall was not a sign of strength but of weakness, built because Rome had become susceptible to invasion by barbarians from the north.

Another threat was internal. The emperor had just quashed a rebellion among a large swath of city workers, and he might have hoped that a massive building project such as a defensive wall would help to employ and placate the masses, while also serving as a show of force to all classes of potentially unruly citizens. On a symbolic level, as a sign of imperial power, Aurelian's wall faced inward as well as outward—for it cast its long shadow on restive Romans and outsiders alike.

Whatever the initial reasons for its construction, Rome's wall has lived an exceptionally long life, but it has hardly remained the same. Some changes have been major—such as the building of the Leonine Wall around the papal enclave in the Vatican during the ninth century—but countless other incremental interventions were necessary to keep the Aurelian Wall standing. Over time, it has been damaged, repaired, expanded, or added to in spots, and updated here and there in response to changing military technology, but like the city itself, it has withstood the test of time. In fact, the Aurelian Wall has been one of the few constants in a place defined by flux. The unceasing attention and upkeep it has demanded and received stands as a barometer to the status of Rome itself: the city as an ongoing process.

During the Middle Ages, the city shrank away from the Aurelian circuit, then, in modern times, burst out of it—outgrowing the wall so dramatically that it no longer demarcates Rome from non-Rome as it once did. The wall is also more permeable than it was in the past, and certainly does not serve its original defensive purpose, but its form remains a signature Roman imprint. Like the omnipresent "SPQR" that stands for "Senatus populusque Romanus" (The Senate and People of Rome)—an inscription that adorns Roman manholes, lampposts, and fountains to this day, but has not really been relevant since the end of the Republic in the first century BCE (fig. 10)—the wall has been repurposed as an emblem of the Eternal City.

Rome before Rome

Published in 1820 as part of a larger volume on the city's underlying terrain and geology, Giovanni Battista Brocchi's map (fig. 11) depicts Rome's very ground and topography as it was "in the early days of the city's foundation." Brocchi was attempting a pioneering *geoarchaeological* study: a reconstruction of Rome's surface features (altitudes, plains, marshes, slopes, etc.) at the time of its mythical origins in the eighth century BCE. Using the highly accurate measurements executed by Giambattista Nolli for his plan of 1748 as a point of departure, Brocchi then extrapolated backward in time by incorporating his own geological samples and observations on what lay beneath Rome's accumulated layers.

Why were such calculations even necessary? We tend to think of natural features like hills and valleys as being relatively immutable—or only changing in "geological" time, over hundreds of thousands, even millions of years—but in fact Rome's ground level has risen considerably in less than three millennia, in some places by dozens of feet. This dramatic change is due to a combination of natural phenomena, like silting and erosion, and manmade factors, like the Roman habit of building new constructions on top of earlier ones, or creating huge trash heaps—like Monte Testaccio, a mountain of discarded amphorae (oil jars)—that eventually became lasting parts of the urban topography. Brocchi's image depicts the physical setting for the city's long history of human settlement, but even that setting was not fixed, for it evolved considerably over the centuries.

That said, Brocchi's image neatly encapsulates Rome on the verge of becoming itself. Before there was a city called Rome, there was a place: a stage waiting for its actors and script. Books have been written on this locale, as if to make sense of how such an unremarkable site could give rise to such a remarkable history. At its heart is a bend in the Tiber River, a serpentine curve, at the hinge of which the river widens, making room for a small island believed to have been the city's first site of human habitation. Nestled within the northern loop of the

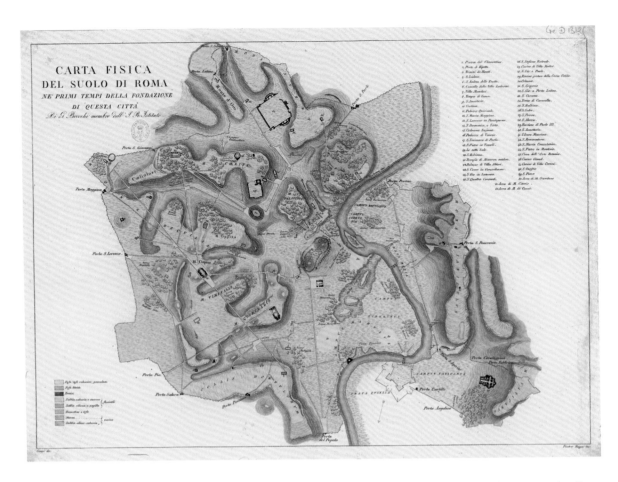

CARTA FISICA
DEL SUOLO DI ROMA
NE' PRIMI TEMPI DELLA FONDAZIONE
DI QUESTA CITTÀ

Fig. 11

Giovanni Battista Brocchi,
*Carta fisica del suolo di
Roma ne' primi tempi della
fondazione di questa città,*
Rome, 1820. Bibliothèque
nationale de France, Paris.

Tiber's curve is a low-lying floodplain, which later became known as the Campus Martius—Campo Marzio in Italian, or Field of Mars. This zone appears
toward the bottom of Brocchi's map, oriented as it is with north at the bottom.
Above it, on the other side of the river, the southern loop contains the zone
that would later become known as Transtiberim (literally, "across the Tiber";
today's bustling Trastevere).

Radiating out from the river are Rome's hills. Most of those on the east side
of the Tiber—at left in Brocchi's map—originated as one large plateau perched
above the river valley, which over time eroded into separate ridges. Among them
are the relatively small but steep hills later known as the Palatine and Capitoline,
which lie right next to each other and very close to the turn in the river's curves,
above its left bank, not far from the Tiber Island. Separated by a valley, these
hills became the core of the ancient city. South of them, equally close to the river,
is the larger Aventine, another site of early settlement. Further from the river, to
the northeast (lower left on Brocchi's map), sprawls another trio of hills made
up of the Quirinal and the Esquiline with the smaller Viminal perched between
them. Above this trio, just southeast of the Palatine, is the Caelian. This array
became the seven original hills of the city.

Rome eventually came to enfold several more hills that lay further from
the Tiber, and these too appear on Brocchi's map. At lower left is the Pincian,

Chapter One

which Brocchi calls "Collis Hortulorum" (Hill of Gardens), as it was known in antiquity. On the right bank of the Tiber, two additional hills bracket the map: the Vatican below and the taller, expansive Janiculum ridge running north-south above.

For anyone who knows Rome even a little bit, it is all but impossible to look at a blank stage like Brocchi's without mentally filling in the later additions. The Pincian Hill is the site of today's Villa Borghese park; the Vatican, world capital of Catholicism; the valley between the Capitoline, Palatine, and Esquiline, home to the Roman Forum; and so on. If you look closely at Brocchi's map, you will see that he too could not resist the impulse to insert later history, for he portrays a handful of Renaissance streets and landmarks, ancient monuments, and churches from the Middle Ages on (fig. 12). There are also later, manmade topographical features—such as the "Agger Servii Tulli" at lower left, which were embankments created for the Servian Wall in the fourth century BCE—surface elements decidedly not present at the time of the city's legendary founding.

Most notably, the map employs the Aurelian Wall as a framing device, using it to hem in the whole image, thereby defining Rome's artificial boundary as if it had been there all along: Rome's version of Manifest Destiny. Brocchi clearly included the wall, like the other anachronistic features, as a key reference point. After all, without its distinctive enclosure, the map could almost be mistaken for any arbitrary assemblage of hills, plains, and river. The lightly outlined

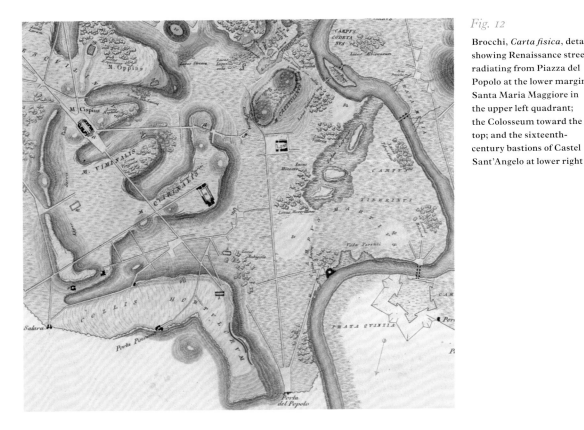

Fig. 12

Brocchi, *Carta fisica*, detail showing Renaissance streets radiating from Piazza del Popolo at the lower margin; Santa Maria Maggiore in the upper left quadrant; the Colosseum toward the top; and the sixteenth-century bastions of Castel Sant'Angelo at lower right

Rome Takes Shape

presence of the Aurelian Wall signals that this is not just a random slice of topography: it is Rome.

A Walled City

This modest map (fig. 13) from the late 1500s captures, in miniature, the central role walls have played in Rome's history from the time of the city's fabled origins. Designed by Dutch geographer Paulus Merula and based, in part, on an earlier map by the antiquarian Pirro Ligorio, the map differentiates layers of Roman history by nesting one inside the other. The city's chronological existence radiates out from the center, with succeeding eras signaled by concentric sets of walls. Merula strips most other elements away—there is scant topography, few streets, and just a selection of monuments—so that the city's long existence is told almost solely through horizontal walled expansion.

The map is also an exercise in ingenious, more-or-less informed speculation. Because Merula made his map during the Renaissance, he was operating at a distance of many centuries—even millennia—from the time span he purported to depict. In his day, scholars were fixated on antiquity in general and on Rome's ancient cityscape in particular, but the modern science of archaeology was in its infancy, and evidence was often lacking. Where physical traces such as ruins were not available to study, scholars could sometimes consult classical literary references to help them figure out what was where, when. Failing that, they had no choice but to rely on their own imaginations to make mapped reconstructions of the city as it had supposedly appeared in the distant past. This scholarly and creative challenge, at once frustrating and stimulating, is discussed further in chapter 5.

Merula's map, meanwhile, embodies both the pitfalls and possibilities of a creative approach. The image is oriented with north at left—the most commonly seen orientation in maps of Rome before north became the gold standard for cartography in the eighteenth century. At the map's core is a rectangular circuit that denotes Rome's earliest form. Because there were no known remains of Romulus's wall, Merula and others struggled to reconstruct its form and extent. All they had to go on were vague references in ancient texts. In a sense, the scarcity of concrete data left room for their imagination to intervene.

Merula's vision of Romulus's city is a bit more substantial than most. Where most later archaeological plans depict *Roma Quadrata* corresponding to the crest of the Palatine, Merula shows the walls encircling its base, creeping down to enfold the valley just to the south—which separated Romulus's settlement from the Aventine and later housed the Circus Maximus, Rome's largest mass entertainment venue—as well as part of the plain that would become the Roman Forum to the north.

Fewer relics of the Servian Wall were known in the Renaissance than today, but Merula was familiar with some material traces of it. He depicts Rome's second wall as a rounded form billowing out from the earlier core. Here, too, he diverges from what would later become conventional wisdom. He smooths the wall out considerably, so it is far from the zigzagging oblong shape that archaeologists now believe to have existed. He also does not depict it as a closed circuit occupying the west side of the Tiber, but rather as an open loop terminating at two distant points along the river. It is unclear whether he meant to depict the wall as being closed in on its fourth side by the Tiber itself, or by another section of wall he represents on the other side of the river, enfolding a large section of Trastevere. There was known to have been an extension to the Servian Wall built on that side of the Tiber, but the stretch Merula depicts corresponds to Aurelian's later circuit. Overall, then, Merula betrays a bit of confusion, fudging the evidence a bit, which is only natural given the information at his disposal.

Finally, Merula does not show the Servian Wall enfolding all seven of Rome's original hills. Large stretches of the Viminal and Esquiline lie outside the walled perimeter, as do some ancient monuments, like the Baths of Diocletian, that probably were constructed within its precincts (albeit much later: a peculiarity of Merula's map is that it conflates eras of history, so that monuments appear together that never coexisted). As if in compensation, he shows other monuments

Fig. 13

Paulus Merula, *Descriptio urbis Romae quadratae et postea . . .* , engraving, Rome, ca. 1594. Special Collections Research Center, University of Chicago Library.

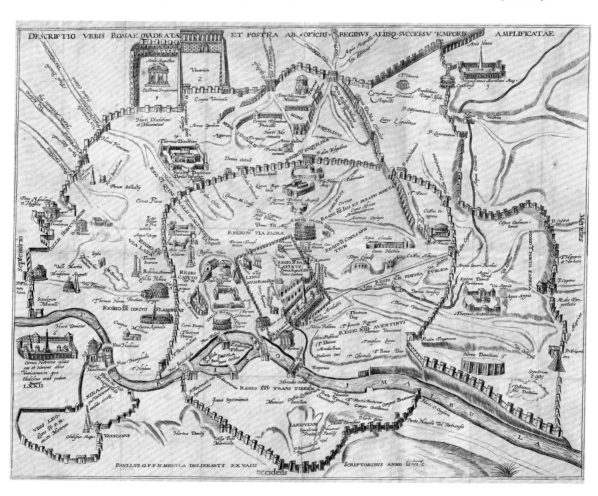

Fig. 14

Opposite: Heinrich Kiepert, *Roma urbs*, from *Atlas antiquus: Twelve Maps of the Ancient World for Schools and Colleges* (Berlin: Dietrich Reimer, 1876). David Rumsey Map Collection, www. davidrumsey.com.

lying inside the Servian Wall that were, in fact, outside of it, such as the Theater of Pompey—again, a later structure, but one with a securely known location.

On Merula's map, the Aurelian circuit defines the outer limit of the city's ancient growth. By his time, even this "new" wall was 1,300 years old (and Rome had actually *shrunk* considerably in the intervening centuries). Merula also added the fortified spur encircling the Vatican, constructed in the ninth century by Pope Leo IV, at lower left—although within its precincts, curiously, there is no sign of St. Peter's, which was the whole motivation behind Leo's defenses. Other than that addition to the Aurelian Wall, none of the circuits Merula represents had ever stood side by side in quite this way. It is a timeless Rome, where everything is present simultaneously.

Merula compresses an amazing amount of urban history into a small, two-dimensional space: just nine by twelve inches. Clearly, the image was meant as a radically abridged version of Rome's development for scholarly viewers who were "in the know"—able to fill in the gaps with considerable outside knowledge. It was not a map for the uninitiated.

Urban Districting

Intended for students, this late nineteenth-century map (fig. 14) presents a clear and straightforward glimpse of Rome's development through subsequent walled stages. Although somewhat dated by the standards of twenty-first-century archaeological knowledge and digital approaches (see, for example, mappingrome.com in chapter 2), it shows how far mapping methods had advanced in the three centuries since Merula. Even if it is inaccurate with regard to some of the details, it effectively conveys how the city was increasingly sliced up into boundaries, visible and invisible, over the course of its early history. In this sense, it is a vivid illustration of the principle that urban development entails an increasingly complex web of bureaucratic as well as physical structures.

Like Merula, Kiepert was a classical scholar: an important pioneer in the field of historical geography. This map comes from his instructional *Atlas Antiquus: Twelve Maps of the Ancient World for Schools and Colleges*. Published in Berlin in 1876, it was the English translation of a work that had first appeared in German in 1854 and would later also be issued by Rand McNally in Chicago. There were more sophisticated archaeological reconstructions circulating in Kiepert's time, such as the one by Luigi Canina (see chapter 2), but those were more geared toward expert audiences. This map is noteworthy because it shaped popular conceptions of Rome's early urban history via its wide use in educational contexts.

Three maps in one, the image separates different stages of Rome's mural and geographical development into adjoining quadrants that together picture the city's growth over time. At lower right is Rome during the time of Romulus, with *Roma Quadrata* outlined in green on the Palatine and the Servian Wall surrounding it in a salmon-pink (fig. 15). Less speculative than Merula's map, Kiepert's still shows the urge to complete fragmentary remains by

Tab. IX.

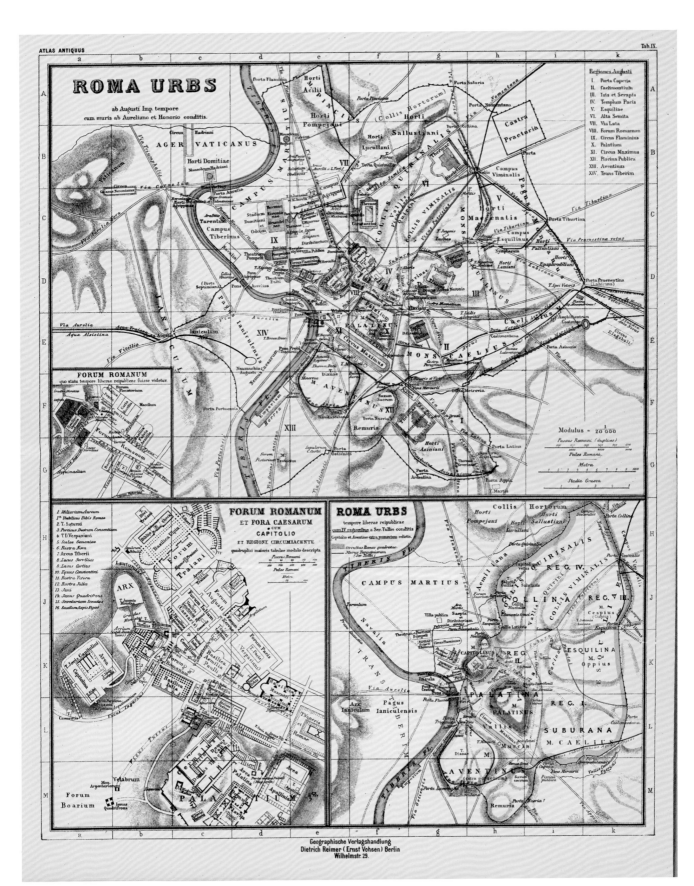

ROMA URBS

ab Augusti Imp. tempore
cum muris ab Aureliano et Honorio conditis.

Regiones Augusti
I. Porta Capena
II. Caelimontium
III. Isis et Serapis
IV. Templum Pacis
V. Esquiliae
VI. Alta Semita
VII. Via Lata
VIII. Forum Romanum
IX. Circus Flaminius
X. Palatium
XI. Circus Maximus
XII. Piscina Publica
XIII. Aventinus
XIV. Trans Tiberim

FORUM ROMANUM
quo statu tempore liberae reipublicae fuisse videtur.

Modulus = $\frac{1}{20\,000}$

Passus Romani (duplices)

Pedes Romani.

Metra

Stadia Graeca

FORUM ROMANUM
ET FORA CAESARUM
cum
CAPITOLIO
ET REGIONE CIRCUMIACENTE
quadruplici maioris tabulae modulo descripta

1. Miliarium Aureum
2ª. Dubitius Urbis Romae
2. T. Saturni
3. Porticus Deorum Consentium
4. T. D. Vespasiani
5. Scalae Gemoniae
6. Rostra Nova
7. Arcus Tiberii
8. Rostra Servilius
9. Lacus Curtius
10. Equus Constantini
11. Rostra Vetera
12. Rostra Iulia
13. Iani
14. Ianus Quadrifrons
15. Secretarium Senatus
16. Sacellum Iapis Nigri

Passus Romani
Pedes Romani
Metra

ROMA URBS
tempore liberae reipublicae
cum IV regionibus a Ser. Tullio conditis
Capitolio et Aventino extra pomoerium relictis.

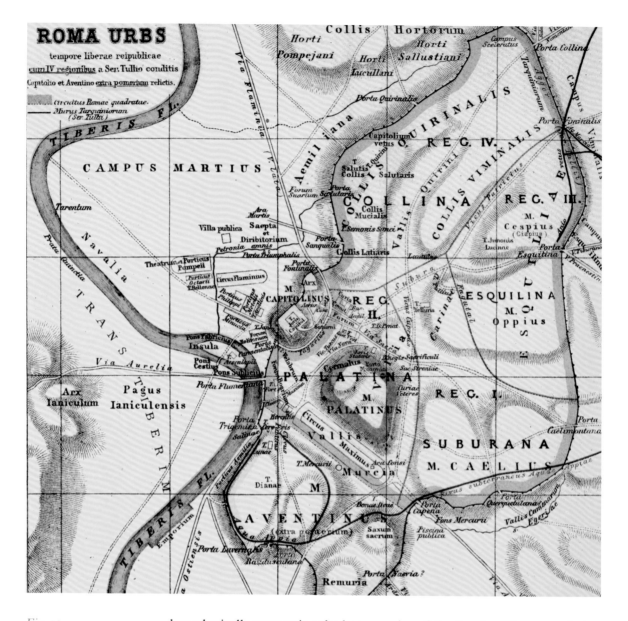

Fig. 15

Kiepert, *Roma urbs*, detail showing *Roma Quadrata* in green, the Servian Wall in pink, and the Capitoline and Aventine Hills in yellow

hypothetically connecting the known ruins of the Servian Wall, completing the circuit.

Servian Rome is shown further carved up into the four administrative *regiones* (or quarters) of the republican period, as indicated by thin red lines. Next to the Servian Wall, the Capitoline and Aventine Hills are marked in yellow to indicate their status as part of the *pomerium*, a buffer zone running around the city wall that marked Rome's religious and territorial boundary.

The map at lower left jumps ahead by several hundred years to give a detailed view of the Roman and Imperial Fora: the zones of daily life, government, and commerce. Just to the south are the palaces of the Palatine; to the west, the temples of the Capitoline: center of the city's religious life. The third map, taking up the top half of the image, shows an ever larger Rome divided into the fourteen

Chapter One

administrative regions of Augustus and encircled by the Aurelian Wall. Here, too, the map freely mixes physical and metaphysical borders to give a picture of the steady accumulations that began to jostle for space on the city's topography.

FURTHER READING

Beard, Mary. *SPQR: A History of Ancient Rome*. New York: Liverlight Publishing Corporation, 2016.

Caputo, Claudio, and Renato Funiciello. "Giovanni Battista Brocchi: La geologia di Roma e la carta del Nolli." In *Roma nel settecento: Immagini e realtà di una capitale attraverso la pianta di G.B. Nolli*, 2 vols., ed. Carlo M. Travaglini and Keti Lelo, 1:43–49. Rome: CROMA—Università degli studi Roma Tre, 2013.

Carandini, Andrea. *Rome: Day One*. Princeton, NJ: Princeton University Press, 2011.

Claridge, Amanda. *Rome: An Oxford Archaeological Guide*. 2nd ed. Oxford: Oxford University Press, 2010.

Coarelli, Filippo. *Rome and Environs: An Archaeological Guide*. Trans. James J. Clauss and Daniel P. Harmon. Berkeley: University of California Press, 2014.

Dey, Hendrik. *The Aurelian Wall and the Refashioning of Imperial Rome, AD 271–855*. Cambridge: Cambridge University Press, 2011.

Funiciello, Renato, and Claudio Caputo. "Giovan Battista Brocchi's Rome: A Pioneering Study in Urban Geology." In *The Origins of Geology in Italy*, ed. Gian Battista Vai and W. Glen E. Caldwell, 199–210. Boulder, CO: Geological Society of America, 2006.

Heiken, Grant, Renato Funiciello, and Donatella De Rita. *The Seven Hills of Rome: A Geological Tour of the Eternal City*. Princeton: Princeton University Press, 2005.

Kiepert, Heinrich, Adriano La Regina, and Richard J. A. Talbert. *Formae orbis antiqui*. Rome: Quasar, 1996.

Kostof, Spiro. *The City Assembled: The Elements of Urban Form through History*. Boston: Little, Brown, 1992.

Zögner, Lothar, ed. *Antike Welten, neue Regionen: Heinrich Kiepert, 1818–1899*. Berlin: Kiepert, 1999.

Chapter Two

Rome of the Caesars

"[Rome] gave ample scope for moralising on the vicissitudes of fortune, which spares neither man nor the proudest of his works, which buries empires and cities in a common grave."

Edward Gibbon, _The Decline and Fall of the Roman Empire_, 1776–89

Under the emperors, Rome experienced a tumultuous half millennium that later eras would look to alternately as a golden age and a cautionary tale. The maps in this chapter show how the city came into its own as a magnetic destination long before our modern tourist era and how the place that had started as a smattering of mud huts on a hilltop above a river reached tentacles out into the world to create a powerful political and administrative network spanning extraordinary distances and diverse cultures. From the visualization of this network in the road map known as the Peutinger Table, to the administrative survey of the urban fabric itself in the marble plan known as the _Forma urbis_ and much later attempts to reconstruct the great capital, in imagery, centuries after its decline, these maps are as much about ideas of empire, and of loss, as they are about a real city and its history.

Rome's conversion from a republic began under Julius Caesar in the first century BCE and then, following a series of power struggles, solidified under his successor, Augustus, who ruled for more than four auspicious decades until his death in 14 CE. Augustus initiated the prosperous and stable period known as the Pax Romana, which lasted two centuries. He restored the system of roads that connected Rome with its far-flung empire and mounted a series of military campaigns that dramatically expanded that empire over large portions of North Africa, northern Europe, the Iberian peninsula, Dalmatia, and Asia Minor.

Augustus transformed Rome into a great metropolis, famously claiming—according to the ancient historian Suetonius—that he had found it a city of brick and left it a city of marble. While that claim might be overblown, he did rehabilitate many older monuments as well as build new ones that are still highlights of the Western canon, such as the Ara Pacis and the Forum of Augustus. Subsequent caesars continued to treat the city as an imperial showcase, creating grand public works that glorified their name while also serving the needs of all Romans. Personal commemoration and public utility were interlocked, together driving the city's growth. The imperial fora are a case in point: Julius Caesar, Augustus, Vespasian, Domitian, and Trajan all laid out large, multipurpose spaces in close proximity to the original Roman Forum.

Some emperors, however, created works of art and architecture that were flagrantly geared toward private pleasure, not public benefit. Nero's opulent, sprawling Golden House (64–68 CE) is the most notorious example. After the great fire of 64 CE had destroyed large stretches of Rome, Nero used the disaster as an opportunity to construct a luxury palace and associated grounds whose footprint took up as much area as many ancient cities. Its features included an

Fig. 16

The octagonal room at the Domus Aurea on the Palatine Hill. Photo: Jonathan Rome, https://romeonrome.com/. CC BY-NC-ND 3.0: https://creativecommons.org/licenses/by-nc-nd/3.0/legalcode.

artificial lake, vineyards, and an octagonal court reported to have been capped by a rotating dome (powered by slaves, naturally) from which rose petals and perfume rained down on guests (fig. 16). That chamber was one of hundreds lavishly decorated with marbles, stuccoes, mosaics, frescoes, and gold leaf. Though that sumptuousness can be challenging to grasp in the stripped-down version that one can visit today, the grandeur of the spaces is undeniable. Even Louis XIV's Versailles would have paled before the extravagance of Nero's Golden House.

After that much despised emperor killed himself in 68, angry citizens condemned the palace as a symbol of decadence and misuse of power. It was entombed in rubble and, in a pointed gesture, used as a foundation for public bath complexes—communal bathing being a ubiquitous activity enjoyed by just about every social class in ancient Rome. The Baths of Titus (79–81 CE) were the first to be constructed on the site, followed later by the much larger ones of Trajan (104–9 CE). The Colosseum, a massive amphitheater similarly meant as a place of entertainment for the people, was built close by and opened with great fanfare in 80 CE.

Engineering was key to Rome's urban growth, just as, later, its decline would be tied to that of the city. Romans exploited the potential of the arch, which allowed them to build taller, longer, stronger bridges and tunnels. They also perfected a recipe for concrete that enabled the construction (and most notably the vaulting) of architectural wonders like the Pantheon, Colosseum, and imperial bath complexes. These and other building technologies allowed Romans to harness the power of water: perhaps the most critical factor in the city's unprecedented growth. Between the fourth century BCE and the third century CE, eleven aqueducts were constructed to bring fresh water into Rome from natural springs originating in the surrounding hills many miles away (fig. 17).

A few aqueducts are visible to this day on the surface of modern Rome, their arches spanning streets and city gates, but most were submerged underground by the time they neared the walls. Within the city, this highly complex arterial network filled Rome's fountains and provided water to private houses while powering public baths, latrines, and sewers. Most aqueducts terminated at the Cloaca Maxima, or "Greatest Sewer," which emptied into the Tiber near the southern tip of the island. Overall, the aqueducts gave Rome a plumbing and sanitation system unmatched until modern times. Needless to say, they required constant upkeep by a veritable army of skilled maintenance workers. Later, when Rome fell on hard times and the aqueducts were neglected, they fell quickly into disrepair—the whole city going along with them, in a steep downward spiral.

The empire peaked in power and extent in the second century, then began to falter over the course of the 200s. For Rome, the construction of the final circuit of walls by Emperor Aurelian was an ominous sign of vulnerability (and, it turns out, an ineffective countermeasure against outside threats). The city did continue to flourish for a time, reaching a pinnacle during the time of Constantine, but its fortunes gradually fell after that emperor left in the 320s to found his "New Rome" at Constantinople. From that point on the empire was split into east and west, which would eventually become divided along linguistic lines— Greek and Latin, respectively. The last emperor to rule over both, Theodosius I, died in 395. Then, in the fifth century, the capital of the Western Empire was moved to Milan, then Ravenna.

Rome was sacked repeatedly—most catastrophically in 410, and again in 455. In 476, the barbarian chieftain Odoacer deposed the last emperor and named himself king: an event traditionally cited as the official end of the Western Empire. Rome limped on, but by the early sixth century its population had dropped by about 90 percent from its first-century high of over one million, and other places had surpassed it in size and influence. A storied chapter in Roman history was coming to a close, even as another was already beginning.

Destination Rome

From its earliest days, Rome did not exist just for Romans. The city's enduring status as a destination is attested to by the city's foremost temple, which was built on the Capitoline in the sixth century BCE and dedicated to Jupiter Optimus Maximus (Jupiter, Best and Greatest). The sheer size of this shrine, which was rebuilt several times and stood amid others on the hill, lived up to that double superlative. It also suggests that Rome was more than a commercial

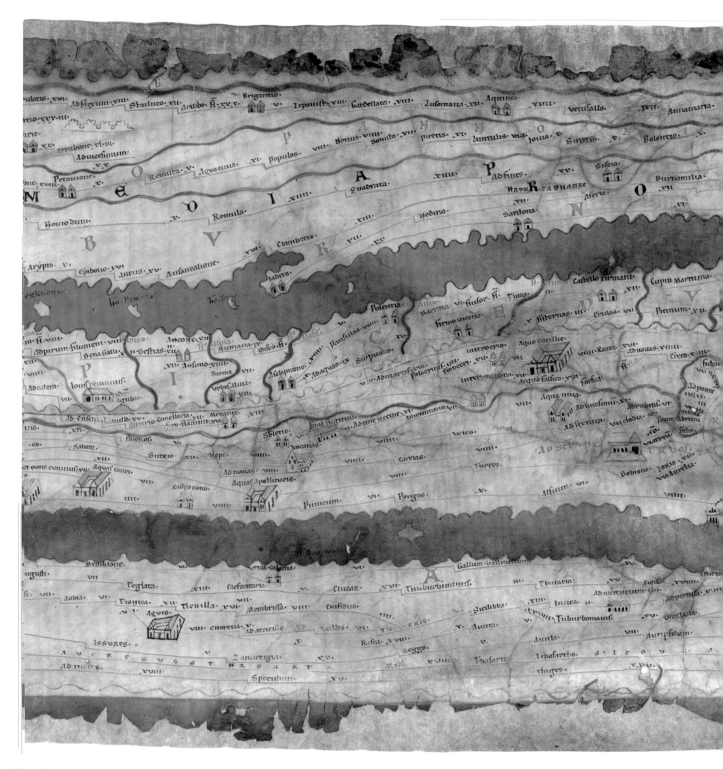

Fig. 18   The Peutinger Table, copy ca. 1200 of untraced fourth-century original. Österreichische Nationalbibliothek, via europeana.eu.

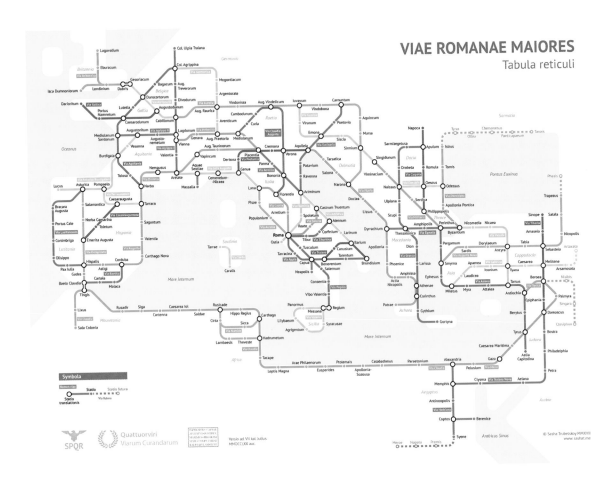

Fig. 19

Sasha Trubetskoy, *Viae Romanae Maiores*, 2017. © Sasha Trubetskoy, https://sashat.me/.

and political hub: it was also an attraction for religious visitors more than a thousand years before the era of Christian pilgrimage. That it remained so for centuries—if ever it ceased to be—is suggested by the remarkable image known as the Peutinger Table (fig. 18).

Named for a sixteenth-century owner, the German humanist Konrad Peutinger, this map is thought to be an early thirteenth-century copy of a fourth-century work—although one scholar has suggested the prototype dates from the ninth century, and still others point to indications it was begun closer to 300 CE and added to over time. For our purposes, suffice it to say that the map encapsulates the Roman world picture of the late antique era. Specifically, it depicts the network of roads that linked the whole empire, centered, of course, on the capital itself.

More than twenty feet long and just over a foot high, the map compresses all of Europe and parts of Asia down to a narrow band, with waterways depicted in green, roads in red. It is often referred to as a strip map for its horizontal format, or alternatively an itinerary map since it represents routes and includes information that would have been helpful to travelers—such as recommended stopping points and the distances between them (even if the unit of measurement is never stated). Like most such maps, however, it privileges the sequence and ordering of places over quantifiable information such as orientation and spatial disposition. In this way, it is comparable to a modern subway diagram (see fig. 19, a clever modern image by cartographer Sasha Trubetskoy that maps the Roman roads through Europe in just this manner).

The purpose of the Peutinger Table, or rather the prototype on which it was based, is as controversial as its dating. Scholars have argued over whether it was intended to serve practical or symbolic ends—as a navigational tool or an ideological expression of Rome's imperial reach. Roman travelers did not use maps like modern travelers do, so the second interpretation seems more likely. But there is no question that the city of Rome was meant to figure prominently on the map, where it appears as the beating heart of the network of roads (fig. 20).

Rome is not depicted as a recognizable cityscape. Instead, it is represented emblematically, as an enthroned, crowned figure holding an orb, symbol of power. The personification also bears a scepter and a shield, its appearance in certain respects akin to depictions of medieval kings (a reminder of the era when the copy was made). It also vaguely resembles the statue of the goddess Roma—a deity embodying the city and its power—that was housed on the Capitoline. Although allegorical, the figure does sit in the midst of a few schematic

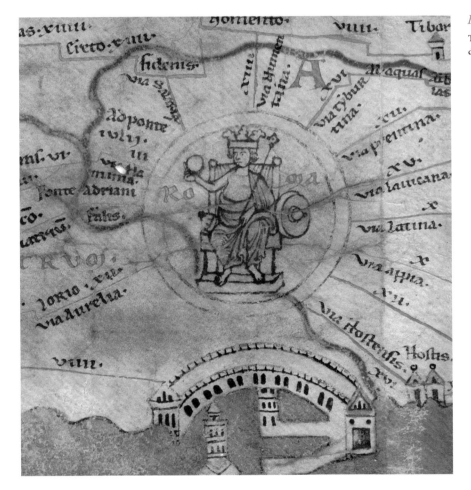

Fig. 20

The Peutinger Table, detail showing Rome

references to Roman topography. Beneath it, on the coast, is the bowed form of Rome's ancient port at Ostia. Encircling the figure, meanwhile, is a yellow ring—traversed by the Tiber—that might refer to the Aurelian Wall, from which many labeled roads radiate outward toward all ends of the known world. Of course, those same roads also converge, coming together at the center of gravity that was Rome itself.

An Incomplete Puzzle

The Peutinger Table offers no answers to a question that has intrigued archaeologists and scholars for at least 500 years: What did the city look like during its glory days at the height of the imperial period? The first attempts at any kind of graphic reconstruction date to the Renaissance, at a distance of a millennium. Over the course of that time much of the ancient city had been buried, despoiled, neglected, and left to quietly fade away. One precious piece of evidence, however, seems to hold some answers about the shape and features of ancient Rome. The Severan marble plan—sometimes called the *Forma urbis*, or shape of the city—is the earliest known map of Rome and the only one to survive from the imperial era (fig. 21).

The problem is that while it is impressively accurate, it is also poignantly fragmentary. We know that the plan was created during the reign of Septimius Severus to adorn a wall of the Temple of Peace in the Forum of Vespasian (fig. 22). It was massive, for it was inscribed on 150 marble slabs that were arranged in eleven rows stretching to forty-three by sixty feet. As with the Peutinger map, there is debate about its function. Some scholars have theorized that the map

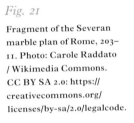

was intended for administrative purposes—essentially a huge taxation survey—while others maintain it was more of a showpiece. In either case, the *Forma urbis* showed Rome in extraordinary detail, with all the city's buildings—from the grandiose to the everyday—as well as city streets and key topographical features depicted in measured ground plan on a scale of about 1:240: a remarkable demonstration of Roman surveying skill. Had the map survived intact, it would solve countless mysteries and controversies regarding Rome and its constituent parts as they existed two thousand years ago.

Alas, it is far from intact (fig. 23). Over the Middle Ages, the map was alternately left to founder and pillaged for construction material, most of its slabs shattered and dispersed through various parts of the city, where they were

Fig. 22

Original setting of the *Forma urbis*, from Roberto Meneghini, *I Fori imperiali e i Mercati di Traiano: Storia e descrizione dei monumenti alla luce degli studi e degli scavi recenti* (Rome: Libreria dello Stato, Istituto poligrafico e Zecca dello Stato, 2009)

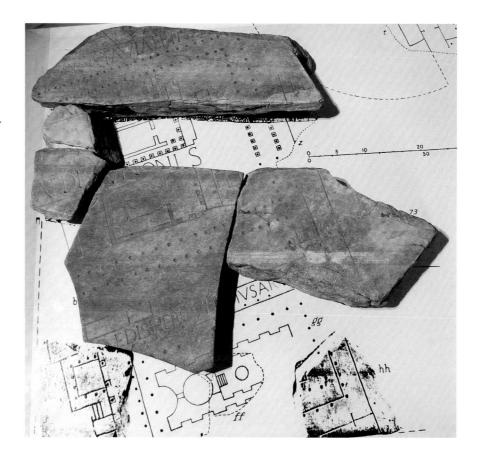

incorporated into new walls and foundations. Only about 10–15 percent of its
original surface area has been recovered to date—the first fragments having
been unearthed in the late 1500s—and that small percentage consists of about
1,200 shattered pieces of marble that scholars struggle to make sense of. Every
so often, another piece turns up, but there is little hope of finding enough for
a substantial reconstruction. Some sections have been cobbled together suf-
ficiently to shed light on individual buildings and neighborhoods, but on the
whole the map is a tantalizing puzzle: the Humpty Dumpty of the history of
cartography. More than a font of information, the Severan plan has become a
pathos-tinged enigma, a symbol of irretrievable loss with regard to the city's
illustrious antiquity.

Making Sense of the Shattered Past

The unbridgeable historical distance of the marble plan is eloquently expressed
by later visual interpretations that attempt to recover both the broken map and,
in a sense, the city it represents. The great seventeenth-century printmaker
Giovanni Battista Piranesi was fascinated by the marble plan and incorporated
its pieces into a number of his etchings. Three quarters of a century later, the
pioneering archaeologist Luigi Canina was similarly enthralled by the broken

Chapter Two

map, taking pains to reconcile its pieces with the physical evidence he was unearthing concurrently through excavations. Piranesi and Canina were both trained as architects, and it is not surprising that they would have been intrigued by the ancient map's measured representation of the built environment. Their responses to that stimulus, however, diverged sharply.

Piranesi's 1756 map of ancient Rome (fig. 24) shows a selection of marble fragments inscribed with monuments that he believed he could identify, arranged around an illusionistic slab of marble cut in the shape of a plan of modern Rome. The outline of the city walls was cribbed directly from Giambattista Nolli's plan of 1748 (see chapter 7), but the interior of the city is almost as bare as Giovanni Battista Brocchi's geological plan, discussed in chapter 1. There are lone bits of ruins here and there, specifically those that Piranesi knew and could still see in the Rome of his day, but he refrains from reconstruction (this is uncharacteristic, as we shall see in chapter 5). Instead, Piranesi simply strips away all later building layers to show what little remained from antiquity. As an archaeological map of the city, the image is left largely blank, as if awaiting further data. In other words, the Severan fragments—as evidence for the form

Fig. 24

Giovanni Battista Piranesi, *Pianta di Roma disegnata colla situazione di tutti i Monumenti antichi . . .* , etching, Rome, 1756. Metropolitan Museum of Art, Gift of Georgiana W. Sargent, in memory of John Osborne Sargent, 1924.

Rome of the Caesars

of the ancient city—have failed to fill the gaps in Piranesi's knowledge, and in his map they are reduced to little more than an artistic framing device.

Almost a century later, Canina took a less romantic approach to the city and the marble plan, treating his *Topographical Plan of Ancient Rome* more like a scientific illustration than a creative expression (fig. 25). Canina, like Piranesi, depicted some of the Severan fragments around the edges of his map, but he neatly spaced them for clarity, not dramatic effect. Gone are Piranesi's artfully haphazard and convincingly hefty marble pieces, his steep contrasts of shadow and light, his dazzling illusionism and etching technique.

Canina did not hesitate to reconstruct ruins, but he did so based on archaeological evidence as well as informed speculation about their original forms. In his map, he depicted Rome's modern urban fabric in light gray silhouette composed of closely parallel-hatched lines. Upon that base, Canina superimposed the ancient monuments, representing buildings (and parts thereof) that still existed in heavy black lines, inferred reconstructions in thinner lines (fig. 26). This graphic technique allowed viewers to grasp instantly the spatial relationship between the city they knew and the city of the past, and to distinguish

Fig. 25

Luigi Canina, *Pianta topografica di Roma antica*, Rome, 1850. David Rumsey Map Collection, www. davidrumsey.com.

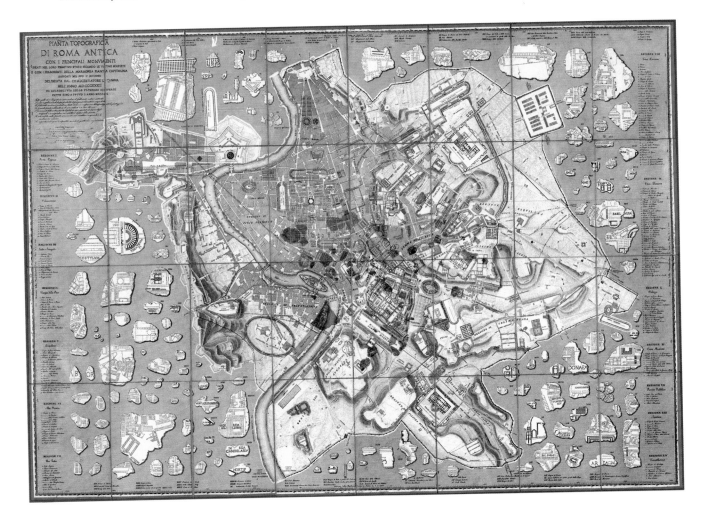

Chapter Two

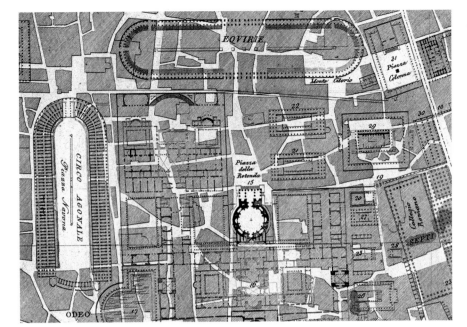

Fig. 26

Canina, *Pianta topografica*, detail showing the area around the Pantheon (center) and Piazza Navona (left)

concrete knowledge from supposition. Canina also signaled on the city plan what information he had gleaned from the *Forma urbis*, for he depicted in shadows the areas that he had managed to correlate with some of its fragments. Generally speaking, his map might lack the visual impact and sentiment of Piranesi's, but it is an informative, ingenious, and cerebral image of the city past and present.

Filling in the Gaps

From 1893 to 1901, Rodolfo Amedeo Lanciani, one of Rome's most venerable archaeologists, published his magnum opus: a graphic reconstruction of the city in meticulous ground plan (fig. 27). He named it the *Forma urbis Romae*, in a direct and deliberate reference to its late antique predecessor, which also took the form of a horizontal footprint of the entire built fabric. Lanciani's plan was, in a sense, his attempt to make good on a promise that that broken relic left unfulfilled: to recreate as complete a picture of the ancient city as possible. The map consists of forty-six plates containing detailed representations of individual sections of Rome, rendered at a scale of 1:1000 (fig. 28). When assembled, the overall image measures seventeen by twenty-four feet and provides a compendium of all the archaeological information known to Lanciani—a vast and erudite array of textual and physical evidence, including identifiable fragments of the Severan marble plan as well as the latest excavation reports, combined in a holistic historical topography.

Yet Lanciani did not stop with the ancient city, for his goal was a comprehensive picture of Rome's development over the centuries, from antiquity through the Middle Ages and Renaissance, up to and including Lanciani's own time.

Rome of the Caesars

This was an intense period—the decades just after Rome had been made capital and was in the midst of radical transformation—and the establishment of *Roma capitale* was an important motivation for Lanciani. He began compiling his labor-intensive project in the 1870s with the express hope that such a sweeping portrait of Rome's entire urban fabric would facilitate a thoughtful, informed approach to the many interventions that were already being implemented—as well as others that were certain to come. And come they did, as we shall see in chapters 9 and 10—but rarely in the historically sensitive and considered manner Lanciani might have wished.

Lanciani took some cues from Canina's map, with its layering of eras and graphic codes, but he improved on those techniques. Canina had simplistically divided Rome into ancient and modern. Lanciani, by contrast, devised a clever and straightforward system of color coding to differentiate buildings and infrastructure from successive periods—using dark red for republican-era monuments, black for imperial through about the year 1000, light red for that time until 1871, and light blue for everything planned or in progress since then. By layering historical strata one over the other, Lanciani managed to condense a remarkable chronological range into a single two-dimensional image. His was a brilliantly simple solution to a problem that had long vexed mapmakers faced

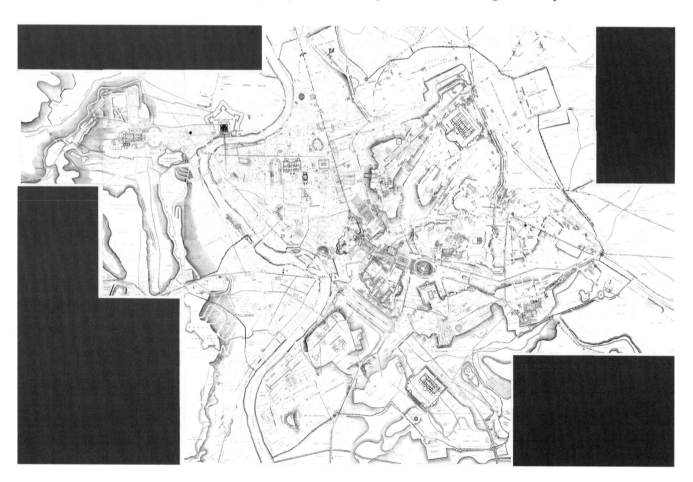

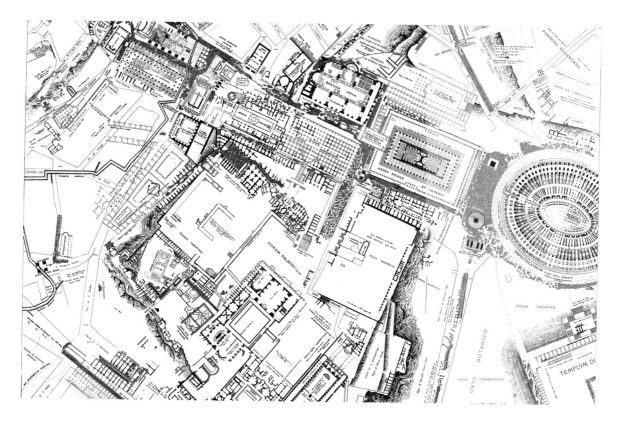

Fig. 28

Lanciani, *Forma urbis
Romae*, detail: sheet 29,
showing the area of the
Colosseum (right), Roman
and imperial fora (above),
and Palatine (below)

with Rome's temporal complexities: the fact that the city does not just exist horizontally across space, but also vertically in time.

Lanciani's *Forma urbis Romae* was far from perfect. More recent scholars have found it to contain inaccuracies and misconceptions, but these are due to inevitable limits of knowledge in Lanciani's time, not flaws in his framework. Lanciani's map is still a foundation and source of inspiration for the most technologically advanced approaches to studying Rome's archaeological makeup over time. In fact, a team of experts from Stanford University, the University of Oregon, and Dartmouth College is currently collaborating on an ambitious project to digitize and remaster Lanciani's *Forma urbis* by "creating a layered, vector version of the map while carefully maintaining the graphic integrity and symbology of the original" (http://www.mappingrome.com/NFUR/). This project affirms the ongoing relevance of Lanciani's achievement, a staying power not unlike that of Rome itself—while updating it with technology Lanciani never dreamed of.

A Model City

One of Rome's most neglected museums is the Museo della Civiltà Romana (Museum of Roman Civilization), located in the former Fascist utopia south of the city known as EUR. The acronym stands for Esposizione Universale di Roma—Universal Exhibition of Rome—because the site, now an office

park, was originally intended to host a 1942 world's fair that never materialized, thanks to World War II. As of this writing, the museum has been shuttered for several years due to budget cuts or planned renovations, depending on whom you ask, but every now and then rumors surface that it will be reopening. Its main attraction, surely, is that it houses one of the most remarkable representations of Rome from any period: the vast three-dimensional model of the Constantinian city known familiarly as the *Plastico* (fig. 29).

This vivid recreation of the imperial city was designed by architect Italo Gismondi and executed by model-maker Pierino Di Carlo beginning in 1932. The motivation for its creation was another Fascist exhibition—the Mostra Augustea della Romanità (or Augustan Exhibition of Roman Culture), billed by Mussolini and his culture ministers as a celebration of the bimillennium of Augustus's birth and held in 1937–38. The *Plastico* was based on the Severan marble plan, even duplicating its scale of 1:250, as well as on Lanciani's *Forma urbis*, although Gismondi incorporated subsequent archaeological discoveries and undertook considerable additional research to raise the city's buildings from flat ground plans to fully realized, volumetric simulations (fig. 30). Gismondi also took the liberty of inventing countless generic "infill" structures to grant a more lifelike picture of Rome's densely settled imperial cityscape. Work on the *Plastico* did not end with the Augustan Exhibition, or with the Fascist regime. Transcending politics, Gismondi and Di Carlo's joint labor of love continued on and off for decades, as the duo continued to improve, update, and tinker with their model into the 1970s.

The *Plastico* presents a beguiling picture of Rome at its peak. From the lavish imperial residences on the Palatine and the temples on the Capitoline to greatest hits like the Colosseum, Pantheon, and Forum, Rome's monuments are intact and glorious, their proportions often exaggerated for legibility. The city's aqueducts are practically flowing, and no corner is left empty. It is a visually exciting, appropriately frenetic, and completely convincing picture. At the same time, when considered in light of the time and circumstances of its creation, it becomes clear that the *Plastico* is about much more than archaeological reconstruction. Viewers in Fascist Rome were meant to draw numerous flattering parallels—to see Mussolini as the new Augustus and Il Duce's Rome as a new golden age.

Whatever its ideological impetus, the model—even more than Lanciani's *Forma urbis*—has taken on new life in the digital age as a basis for computerized reconstruction. The Museum of Roman Civilization purportedly has plans for a digitized version linked to informative metadata for the major buildings, and the *Plastico* has also been an important reference point for Bernard Frischer's Rome Reborn (fig. 31), "an international initiative whose goal is the creation of 3D digital models illustrating the urban development of ancient Rome from the first settlement in the late Bronze Age (ca. 1000 B.C.) to the depopulation of the city in the early Middle Ages (ca. A.D. 550)."

In the latest version, Rome Reborn 3.0, Frischer has departed from Gismondi's model and its analog limitations to create an immersive recreation. With the assistance of a virtual reality headset, people can enter into a breathtakingly evocative simulation of the ancient city, glimpsing it from above in flyover mode or exploring it from street level in a way Gismondi's model simply does not permit. Yet even the most advanced digital technology has not completely dimmed the power of Gismondi's painstakingly handmade model, which still exercises considerable allure. When the Museum of Roman Civilization is open to the public, visitors are able to walk above and around Constantine's city on an

Rome of the Caesars

Fig. 31 Screenshots from Bernard Frischer's Rome Reborn showing the west side of the Roman Forum and the city from above, https://www.romereborn.org. © Flyover Zone Productions.

elevated viewing platform, savoring every detail as if from an airplane—albeit one that doubles as a time machine. So go see it (but call ahead to make sure it's open).

FURTHER READING

Albu, Emily. "Imperial Geography and the Medieval Peutinger Map." *Imago Mundi* 57 (2005): 136–48.

Meneghini, Roberto, and Riccardo Santangeli Valenzani, eds. *Forma Urbis Romae: Nuovi frammenti di piante marmoree dallo scavo dei Fori Imperiali.* Rome: "L'Erma" di Bretschneider, 2006.

Pavia, Carlo. *Roma antica, com'era: Storia e tecnica costruttiva del grande plastico dell'Urbe nel Museo della Civiltà Romana.* Rome: Gangemi, 2006.

Reynolds, David West. "Forma Urbis Romae: The Severan Marble Plan and the Urban Form of Ancient Rome." PhD dissertation, University of Michigan, 1996.

Rodríguez-Almeida, Emilio. *Forma urbis marmorea: Aggiornamento generale 1980.* Rome: École française de Rome, 1981.

Talbert, Richard J. A. "Rome's Marble Plan and Peutinger's Map: Continuity in Cartographic Design." In *"Eine ganz normale lnschrift" . . . und ahnliches zum Geburtstag von Ekkehard Weber: Festschrift zum 30. April 2005,* 627–34. Vienna: Österreichische Gesellschaft für Archäologie, 2005.

Talbert, Richard J. A. *Rome's World: The Peutinger Map Reconsidered.* Cambridge: Cambridge University Press, 2010.

Talbert, Richard J. A., and Richard W. Unger, eds. *Cartography in Antiquity and the Middle Ages: Fresh Perspectives, New Methods.* Leiden: Brill, 2008.

Tschudi, Victor Plahte. "Plaster Empires: Italo Gismondi's Model of Rome." *Journal of the Society of Architectural Historians* 71 (2012): 386–403.

Chapter Three

Rome of the Popes

"I thanked God, mighty throughout the entire world, who had here rendered the works of man wondrously and indescribably beautiful. For although all of Rome lies in ruins, nothing intact can be compared to this."

Magister Gregorius, *Marvels of Rome*, twelfth century

As the Roman Empire came grinding to a halt, a Christian order was already on the rise that would breathe new life into the city. Few maps survive from this period, but those that do bear witness to a time of stimulating intellectual and artistic advances, far from the old-fashioned notion of the Dark Ages. During the long Middle Ages—a period lasting, in Rome, from about the fifth to the fifteenth centuries—mapmaking, like culture in general, flourished in monasteries and convents far from cities. It is telling that of the relatively small number of maps known from this period, most were made by clerics and were infused with a Christian perspective on world history and Rome's place in it. Created, for the most part, far away, they tended to see the city through a distant lens and to frame it in allegorical, metaphysical, or poetic terms.

Rome's Christian history is as old as Christianity itself. The apostles Peter and Paul are just two of the scores of martyrs who met their fate in the city, which became a place of shrines, as growing ranks of the faithful came to venerate sites linked to the lives and deaths of holy figures. The emperors were slow to grasp that this religious sect presented more of a challenge to the status quo than others that had cropped up around the Mediterranean.

After a few centuries of sporadic, if severe, persecutions, Constantine yielded to the rising tide by legalizing Christianity in 313—later, he would become its most high-profile convert. Then, in 391, Theodosius I made it the official state religion. These events, coupled with the departure of the imperial court for Constantinople in the 320s, paved the way for the bishop of Rome—an office later known as pope—to become a powerful force in the city and throughout the Western world.

Rome today is a city of churches, but church building in Rome and elsewhere was all but nonexistent during the first three centuries of Christianity, for the simple reason that the religion itself was illegal, so its practitioners were none too keen to advertise their activities. Worship tended to take place furtively, in private houses owned by well-to-do converts. All that changed dramatically in the 300s. The key figure was again Constantine, who sponsored the building of grand structures marking the city's holiest places. It is a touch ironic that all this construction activity coincided with the time he was deserting Rome for his new and rival capital at Constantinople, effectively splitting the empire into east and west.

Be that as it may, Constantine set the tone for Rome's emerging saintly side. In the 320s, the tomb of St. Peter was enshrined within his namesake basilica on the Vatican Hill. Outside city walls and across the river from the bustling part of Rome, this setting was not exactly central. But it was the location of the circus built by Nero in the first century, where Peter had been crucified in 64 CE along with scores of his fellow Christians, in a gruesome but not uncommon Roman punishment-spectacle.

Fig. 32

St. Peter's Basilica, Rome. Photo: Wolfgang Stuck / Wikimedia Commons.

Rome of the Popes

Soon after his death, Peter's followers recovered his body and buried it nearby. That spot quickly became an early Christian memorial. Then, more than 250 years later, Constantine built a great basilica in the apostle's honor, centered on that focal point. Eventually, the high altar was placed directly over Peter's tomb, which remained the centerpiece of the even grander structure that replaced the fourth-century church in the Renaissance (fig. 32). Michelangelo's great dome, the most prominent feature of "New" St. Peter's and of Rome's skyline, rises directly over the apostle's grave, drawing attention to it from miles away. For close to two thousand years, therefore, Peter's remains have beckoned pilgrims as the main attraction of Christian Rome.

Less known and visited today is the burial site of St. Paul, but it shared center stage with Peter's shrine for much of the Middle Ages. The two apostles were the city's dual patron saints, considered the successors to Romulus and Remus as founders of the Christian city: a special status that is still recognized in the celebration of their joint feast day on June 29 (fig. 33). Like Peter, Paul was martyred on the outskirts of the city in retribution for his missionary activities—but as a Roman citizen, he was granted the favor of a quicker death by decapitation. His tomb, located outside city walls along the Ostian road south of the city, received similar treatment to Peter's during and shortly after the time of Constantine. The first building marking the spot was a relatively modest church, replaced just a few decades later by an immense basilica that lasted for almost 1,500 years, until it burned catastrophically in 1823 (the structure there today is largely a nineteenth-century reconstruction incorporating a few surviving parts of the original).

Constantine also funded St. John Lateran, generally considered the first of the great Christian basilicas in Rome. Originally dedicated to Jesus Christ himself, it was later rededicated to both St. John the Baptist and St. John the Evangelist. If the Lateran Basilica did not rise over a saint's tomb, like the churches of Peter and Paul, it had the distinction of being Rome's cathedral and the official seat of the papacy. Over the following three centuries, many other venerable basilicas followed, including Santa Maria Maggiore in the fifth century and San Lorenzo in the sixth. In this way, the city's sacred topography slowly took shape, attracting more and more pilgrims.

Not all churches were new constructions. Some pagan buildings were repurposed for Christian use, the most famous being the Pantheon (fig. 34). Built by the Emperor Hadrian in the second century on the site of a previous temple erected by Marcus Agrippa that had burned to the ground, this awe-inspiring domed structure was dedicated to all the pagan gods. In 609, it was converted into a church by Pope Boniface IV, who thereby claimed it in the name of the one and only God.

In fact, many of the ancient buildings that survive in relatively good shape today owe their condition to the new functions and identities they assumed during the Middle Ages. A second example is the second-century Mausoleum of Hadrian, another unmistakable feature of the city's skyline that serves as a gateway of sorts for the Vatican (fig. 35). This hulking, cylindrical imperial monument was converted into a fortification shortly after 400, then came to be venerated as a Christian miracle site after Pope Gregory the Great had a vision, in 590, of the Archangel Gabriel hovering above it as he sheathed his sword—in a prophetic sign that the plague that had been raging through Rome was about to cease. It is to this legend that the structure owes its modern name, Castel Sant'Angelo.

The building changed hands frequently over the following centuries, occupied by a series of powerful, feuding Roman families, then was transformed into a major papal stronghold beginning in the 1300s. It has since served as a palace, a prison, and—most recently—a museum. Such examples remind us again and again that Rome is synonymous with reuse and reinvention.

At the same time, the early Middle Ages witnessed Rome's continuing urban decline. The sixth century was a particularly low point, as the city was caught in the crossfire of warring forces. On one side was the Ostrogothic Kingdom, comprised of a Germanic tribe that migrated south from Central Europe to dominate the Italian peninsula in the late 400s, on the other the Eastern Roman (or Byzantine) Empire, composed of Greek-speaking Christians whose

Fig. 34

The Pantheon, Rome.
Photo: Manuel Cohen / Art Resource, NY.

emperor—following in Constantine's footsteps—ruled from Constantinople. Both factions established their capital in Italy at Ravenna, a city 200 miles north of Rome, which was thus deprived of its former political centrality—but it was still a symbolic prize worth fighting over.

Besieged and brutally sacked, Rome passed back and forth between these two powers repeatedly in the 530s and 540s. By the time hostilities concluded in the 550s, the city was in shambles. Aqueducts were in disrepair and the Tiber embankments were crumbling. The city's population fell to fewer than 50,000 and contracted toward the river—the readiest source of water. Other parts of Rome, particularly its eastern and southern reaches, grew increasingly desolate, inhabited mainly by grand, decaying ruins: silent witnesses to a once flourishing city. Conditions were not to improve substantially for nearly a thousand years.

Even as Rome shrank to a shadow of its former self, the popes became more powerful as de facto rulers of the city. The papacy was and is a distinctly Roman institution, a nonhereditary monarchy that wielded political as well as religious power from the Middle Ages until modern times. Fundamental to papal legitimacy is the belief that each successive pope traces his authority back to Peter and ultimately to Christ, who uttered the words: "thou art Peter, and upon this rock I will build my church . . . And I will give unto thee the keys of the kingdom of heaven: and whatsoever thou shalt bind on earth shall be bound in heaven: and whatsoever thou shalt loose on earth shall be loosed in heaven" (Matthew 16:17–18). Throughout its history, the papacy has hinged on this message of timeless authority bestowed by Christ and of the pope as intercessor between God and humanity.

The earliest popes are a hazy group, but beginning in the time of Constantine, the historical record becomes a bit clearer. Gregory the Great was a pivotal and charismatic figure ca. 600, as was Leo III—who sealed an alliance with Holy Roman Emperor Charlemagne—two centuries later. When Charlemagne journeyed to Rome to be crowned by the pope in St. Peter's on Christmas day of the year 800, it was a sign of papal as well as imperial influence. Under the popes, Rome's appeal as a pilgrimage destination also continued to grow unabated. Less arduous to reach than the Holy Land, the city was exceptionally rich in saints' relics and sacred sites. If its fortunes had fallen since late antiquity, its historic and religious cachet were undimmed.

The ninth to twelfth centuries were an unsettled period marked by intermittent conflict with the Holy Roman Emperors and internal power struggles among Rome's noble (or baronial) families, as well as between those families and the popes—who were often members of those same kinship networks. The territorial disputes that took place within city walls were constant and vicious, like gang warfare. Medieval Rome was essentially a hostile environment, the cityscape bristling with private towers that served the practical purpose of defense and the symbolic purpose of advertising power (fig. 36). From time to time, governing authorities would force the towers to be cut down to size, but they had a stubborn tendency to grow back. For centuries these medieval skyscrapers remained Rome's most distinctive and visible feature. Of the multitude that once existed, only a handful—most notably the Torre delle Milizie and Tor de' Conti—remain standing (fig. 37).

By the late Middle Ages, the city seemed to be on the road to recovery, thanks to a prolonged upswing in papal authority and initiatives. In the late 1200s, Pope Nicholas III, a member of the noble Orsini family, focused his renewal efforts on the papal palaces adjoining St. Peter's and the Lateran. A few

Fig. 36

Anonymous, view of Rome, ca. 1538, Museo della Città, Mantua (detail showing city center). Photo: Scala / Ministero per i Beni e le Attività culturali / Art Resource, NY.

years later, in 1300, Boniface VIII instituted the first Jubilee, during which the faithful were encouraged to visit Rome and its sacred sites, lured by the promise of indulgences (that is, remission from certain sins). These Holy Years, which eventually occurred every quarter century, drew huge numbers of pilgrims. Boniface also founded Rome's university, dubbed "La Sapienza," or "holy wisdom," in 1303.

The cultural momentum was short lived, however. The Avignon Papacy—during which the entire papal court decamped to southern France after a power struggle between the French king and the papacy and the forced election of a French pope—lasted from 1309–77, leaving Rome to the mercies of rival clans. The subsequent Great Schism, a chaotic period of competing claims to papal

Chapter Three

authority, lasted until 1417. By that time, after more than a century of disorder, Rome was at its nadir—although yet another resurgence was near.

In sum, stability was elusive in Rome across the long Middle Ages. The city's identity grew more layered and complex, its ancient history and patrimony complemented—arguably surpassed—by an equally impressive Christian heritage. It was a place of piety and paganism, prestige and squalor, growth and decay. How was such complexity expressed in maps? Alas, our ability to answer that question has limits. Maps seem to have been produced for a variety of purposes both symbolic and practical during the Middle Ages. Anecdotally, we know of plentiful local, tax, and property surveys, navigational charts, world maps, and other types of cartography. But just a tiny fraction has survived, and we simply cannot recapture fully the richness of medieval cartography. Still, the images of Rome that have come down to us permit glimpses into medieval perceptions of an evolving, multifaceted, enduring city.

Sacred Buildings and Secular Symbols

Two of the most noteworthy late medieval images of Rome were conceived in the early to mid-1200s by learned clerics active in distant northern Europe—people who had probably never seen the city in person but whose imagined visions reveal a great deal about how Rome was perceived. Matthew Paris was a Benedictine monk at St. Albans Abbey in southern England. His famous "itinerary map," which exists in two versions, shows consecutive towns along the main pilgrimage routes from London to Jerusalem, with Rome as an important stop along the way (fig. 38). The path is indicated by a continuous line linking locations in vertical columns, meant to be read from bottom to top, and left to right.

Like the Peutinger Table, this image is a strip map, albeit one that is laid out across the pages of a volume rather than in one long band. Again, the sequence of places is prioritized over measurable quantities like the distances between them, their orientations relative to each other, or scale. That said, the element of time is built into Matthew Paris's map, for the line connecting any two adjacent stops was meant to equal a day's journey. Each place is labeled with its name and a small, schematic drawing of a wall, tower, or building. The importance assigned to a town is indicated by how elaborately it is depicted, but the illustrations and landmarks are generic. Realism was not a goal.

Not surprisingly, the view of Rome is one of the more elaborate in Matthew Paris's itinerary (fig. 39). Even so, it is a highly simplified version of the city, which would be impossible to identify were it not labeled. Rome is shown as a rectangular walled perimeter with a river winding through it. City gates at left and right respectively lead to "Poille," or Puglia, to the south, and "Lumbarde," or Lombardy, to the north. Otherwise, just four structures are depicted: St. Paul's Outside the Walls; the Church of Domine Quo Vadis, another popular pilgrimage spot, where St. Peter was believed to have had a miraculous encounter with Christ, who chastised him for trying to flee his fate in Rome; St. John

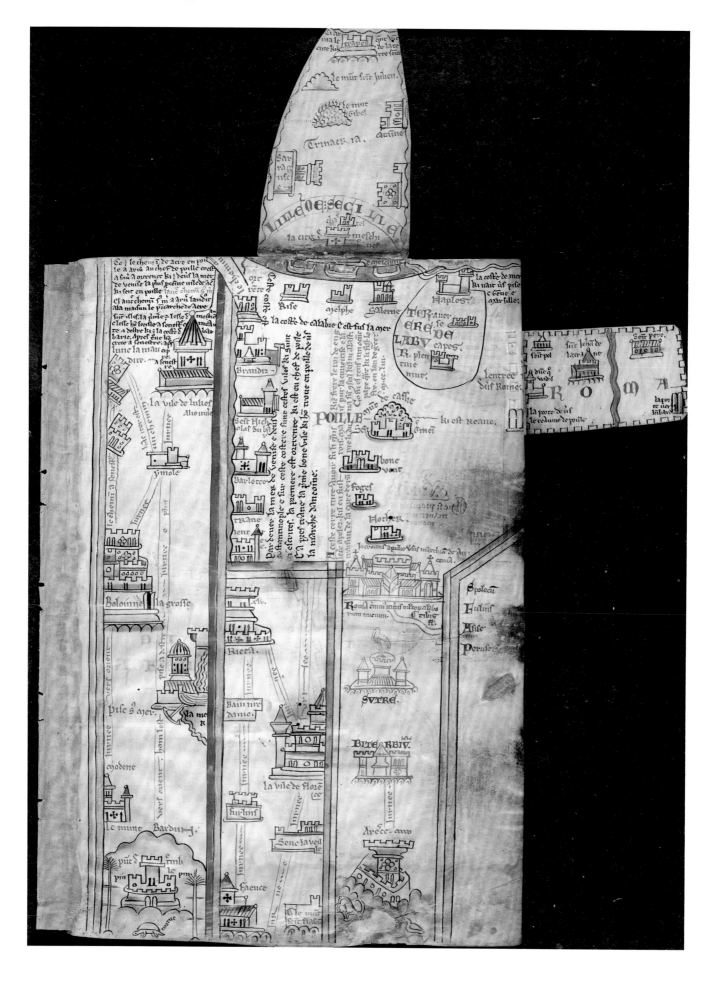

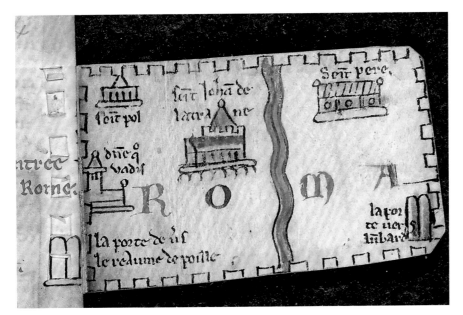

Fig. 38

Opposite: Matthew Paris, *Historia Anglorum*, Royal MS 14 C VI, fol. 4r, ca. 1250. The British Library / Granger.

Fig. 39

Left: Matthew Paris, *Historia Anglorum*, detail showing view of Rome

Lateran; and St. Peter's. The two first sites are situated, incorrectly, within city walls, while St. Peter's is placed, correctly, on the opposite side of the river from the Lateran.

Accuracy, like realism, was not beyond the abilities of the mapmaker: it was beside the point. This map was not meant as a travel aid. Rather, it was meant to substitute for a real religious voyage by enabling *imagined* pilgrimage for monks cloistered in their remote abbey. As such, only a very narrow slice of Rome was necessary, or desired. The city's myths, history, hills, pagan monuments, and streets would have been distractions and are accordingly omitted. Rome is defined by just four Christian landmarks, an enclosing wall, a river—and a place within the larger pilgrimage network.

More identifying features grace the depiction of Rome from the late medieval Ebstorf mappamundi: an enormous, painted world map that was rediscovered at a Benedictine convent in Lower Saxony (modern Germany) in the nineteenth century (fig. 40). Presumably, the map was the magnum opus of a nun (or nuns) in that setting. In the past it was attributed to one Gervase of Tilbury, but serious questions have been raised recently about its authorship, as well as its dating and function. Destroyed by allied bombing during World War II, it is known today only from photographs. While we cannot state with certainty if it was primarily geared toward instruction, devotion, or some other purpose, there is little doubt it was a prestigious item meant for prominent display.

The map pertains to a category known as *mappaemundi*: a Latin term that simply means "maps of the world" but is used to refer to a distinct type of late medieval religious cartography. Such images were almost always created in monastic settings, by monks and occasionally nuns who filled them with a compendium of knowledge about the world. In the Ebstorf mappamundi, as was typical, the earth is shown as a circle, with east—location of the Earthly Paradise or Garden of Eden—in a position of privilege at the top. The three

Rome of the Popes

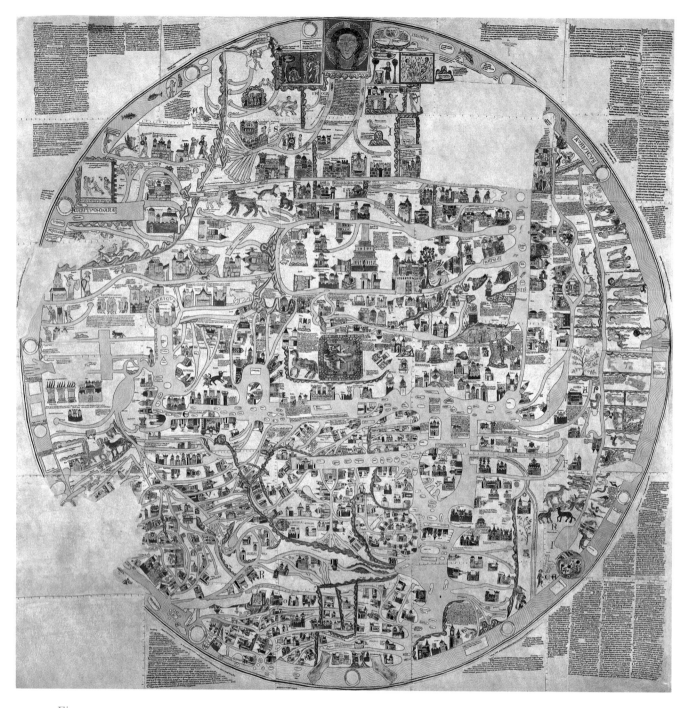

Fig. 40

Ebstorf mappamundi, early
thirteenth century, or ca.
1300, destroyed (formerly
Hanover, Germany)

known continents are compressed within that circular frame: Asia above, Europe below, and Africa at right. Jerusalem is depicted at the center, its square city wall here shown containing a representation of Christ's resurrection.

In the Ebstorf mappamundi, as in some others, the earth is meant to be understood as the body of the crucified Christ. His face appears at the top, depicted like a religious icon, his hands at left and right, and his feet at the bottom. Throughout the map he envelopes, small pictures and labels convey places and episodes from classical myth, popular legend, and biblical history, alongside current sites. Locations tend to be left vague: as with Matthew Paris, spatial precision and pinpoint accuracy were not the goals of the image. The map contains an encyclopedic amount of information, both sacred and secular, but there is little doubt that this is a highly Christianized cartography and world view.

The depiction of Rome is located toward the map's center, about one-third of the way up (fig. 41). Like Matthew Paris, the mapmaker has captioned the city with its name and enclosed it within a simplified circuit of turreted walls—this time in oval form—interrupted by a river crossed by a bridge. Flanking the river are six monuments, including St. Peter's, St. John Lateran, Santa Maria Maggiore, and other important churches. St. Paul's and two other extramural shrines are placed above the city and outside its walls.

Also as in Matthew Paris's map, there is no attempt to individualize these structures beyond the addition of labels. This image, too, is about Rome's essential Christian identity, and little else. There is no sign of the city's ancient patrimony—unless you count the Mausoleum of Hadrian, which by this time had been Christianized as a papal fortress and a place where an angelic miracle had occurred, or the Pantheon, marked "Rotunda," which probably merited inclusion for having been converted into a church, beyond its inherent fascination as an architectural marvel.

But there is something more to this image, for a large roaring lion stands sentry at upper left, resting its rear- and forepaws upon two towers in the city's wall.

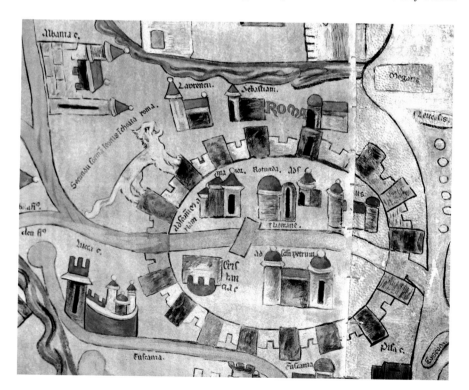

Fig. 41

Ebstorf mappamundi, detail showing view of Rome

It has been suggested that this symbol might be a clue to the map's patronage—namely, that it originated in the court of the Guelph dynasty of Brunswick (Braunschweig) and Lüneberg, Lower Saxony. These staunch papal supporters probably would have been eager to advertise their loyalty to Rome on a magnificent map.

Whether or not that propagandistic element was a factor motivating the appearance of the lion, its presence on the Ebstorf mappamundi held more general significance relating to Rome. An inscription nearby notes, "secundu[m] forma[m] leonis i[n]choata [est] Roma": basically, Rome is in the form of a lion. This statement refers to a common late medieval association of the city with this animal symbol—one that had little to do with Christianity and everything to do with dominion. One twelfth-century scholar had written that Rome ruled over other cities as the lion did over other beasts.

More specifically, the lion was connected to Rome's unsung secular government. In the 1100s, an attempt had been made to set up a senate along the lines of the ancient republican model, in order to counterbalance the pope's power. This body was constantly embattled, and by 1200 it had been whittled down to a single senator who wielded no real power but who "ruled" from a fortified palace on the Capitoline Hill—which thus assumed a new identity as Rome's civic center, distinct from its ancient identity as the city's religious center.

At the foot of the senator's palace stood a statue of a lion attacking a horse, understood as a symbol of sovereignty, strength, and justice. Whether or not the mapmakers were aware of this secular allegory, in the Ebstorf map, they attached its meaning to an otherwise thoroughly sacred vision of Rome. The lion symbol, moreover, with its simultaneous, overlapping associations, hints at the many meanings Rome itself could hold for different constituencies, near and far.

The Medieval Cityscape

There was more to medieval Rome—and to medieval maps of the city—than its Christian shrines. Less than a century after Matthew Paris and the Ebstorf mappamundi, a Franciscan monk and historian known as Fra Paolino Veneto created a very different picture (fig. 42). Active primarily in cosmopolitan Venice, Fra Paolino was a well-traveled bibliophile, a diplomat with friends in high places, and something of a bon vivant. His map of Rome—which was included in a manuscript of his *Chronologia Magna*, a universal history from Creation to the present—was as different from the works of his cloistered northern European counterparts as was (presumably) his urbane background and lifestyle.

While it is quite likely that the others had never laid eyes on Rome, it is fairly certain that Fra Paolino knew the city, and knew it well. Admittedly, his map has its share of mistakes, and his tendency to orient buildings, hills, and inscriptions in different directions can be disconcerting to modern eyes. That said, the map is redolent of firsthand knowledge—and of a desire to convey a full view of the city's natural and manmade features, which are shown with a level of specificity that is exceptional for the time.

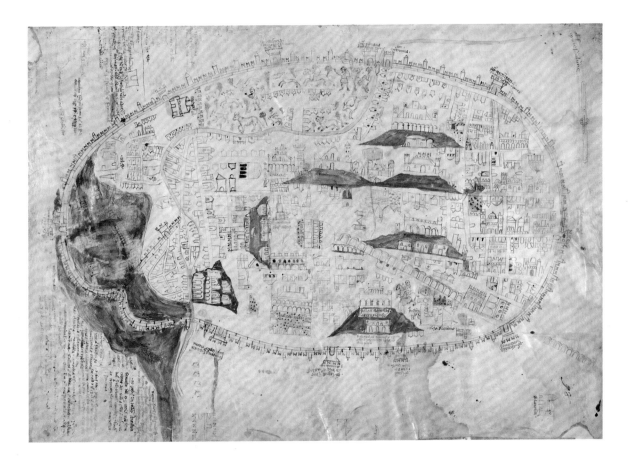

The oval shape of the Aurelian Wall is similar to the smoothed-out circuit of the Ebstorf map, but there the similarities end. Within city walls, Fra Paolino's topography is rendered in basic but recognizable detail—the curving Tiber with its island, the hills—as is the architecture, both sacred and secular, which is shown in abbreviated fashion but often with a nod to actual features. Close inspection reveals St. Peter's at left, flanked by its bell tower and papal palace, and St. John Lateran mirroring it at lower right (fig. 43). The Colosseum is at center, shown domed in accordance with a legend that it had been covered in antiquity, while above and to the left of it is the Pantheon, with its own very real dome (fig. 44). Between them is the Capitoline Hill capped by the Senator's Palace with its tower.

Additional landmarks are easy to identify, with or without labels, due to their appearance and their place in the larger surroundings. The Lateran Basilica is flanked by a sculpture garden, for many ancient statues had been collected there over the centuries, including the famed bronze Marcus Aurelius equestrian statue and the colossal stone head of Constantine. Later, in the fifteenth century, these and other works would be moved to the Capitoline, where they still reside today.

Equally remarkable on Fra Paolino's map is the portrayal of infrastructure, especially the streets that knit the city together. Repetitive, anonymous structures heighten the sense of Rome as an urban fabric, not a collection of emblems.

Fig. 42

Fra Paolino Veneto, view of Rome from *Chronologia Magna*, ca. 1323. By permission of the Ministero per i Beni and le Attività Culturali–Biblioteca Nazionale Marciana, ms Lat. Z. 399 (=1610), fol. 98r.

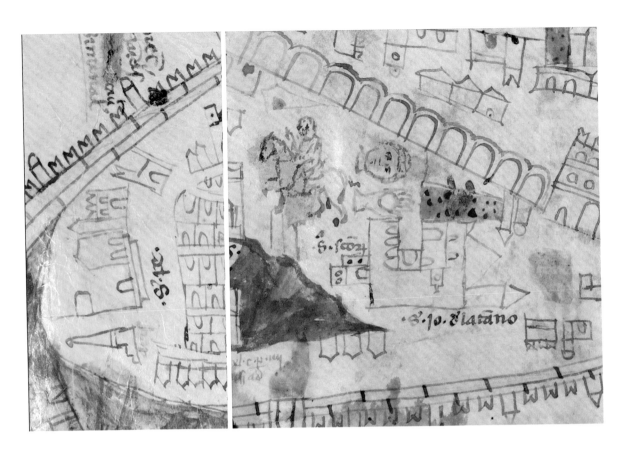

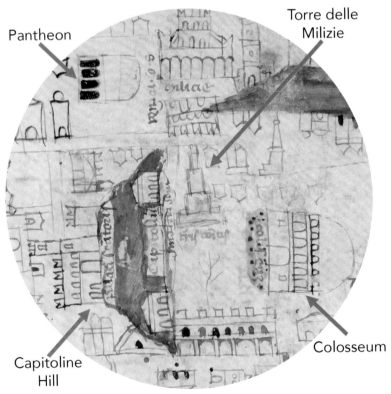

Pantheon

Torre delle Milizie

Capitoline Hill

Colosseum

Towers extend from many fortress-like buildings, symbolizing the clashing powerful families who effectively ruled (or misruled) Rome during this time, when the papacy had abandoned the city for Avignon. Overall, however, Fra Paolino's picture of "popeless" Rome is not one of squalid disorder. Even at its low point, it is a vibrant city defined by landmarks of all stripes, brought to life through a plausible evocation of its real features and layout.

Pathos and Wonder

Fig. 43

Opposite, above: Fra Paolino Veneto, view of Rome from *Chronologia Magna*, detail showing St. Peter's (left) and St. John Lateran (right)

Fig. 44

Opposite, below: Fra Paolino Veneto, view of Rome from *Chronologia Magna*, detail showing city center

Although this small image (fig. 45) was created in the mid-1400s, it was based on a lost prototype dating from a century before and is—like Fra Paolino's map—a window onto the rich variety of late medieval approaches to the city. Rome is oriented with south at the top: the so-called "pilgrim's perspective," because it reflects the point of view of a traveler arriving on foot from the north. The image is more selective than Fra Paolino's, with just a smattering of landmarks scattered across a pared down cityscape, but those landmarks are portrayed in more detail than we have seen, as is the irregular circuit of the Aurelian Wall.

While Fra Paolino's map was fairly neutral and documentary, this one is marked by melancholy and nostalgia for the city's past greatness. It was an illustration for a manuscript of the *Dittamondo*, a travel epic of the mid-1300s based loosely on Dante's Divine Comedy that recounts the fictional world tour of the author, Fazio degli Uberti, with the ancient geographer Solinus as his guide. In this image of their arrival at Rome, the pair appears at upper left, peering over the walls at the cityscape below.

Their gaze falls on an old woman dressed in a heavy black cloak who huddles by the side of the Tiber, next to the Colosseum, and looks meekly back at them. This figure is none other than Rome herself—the city personified as a widow grieving the loss of her rulers: first the caesars, then the popes. How different she is from the enthroned ruler of the Peutinger Table, or the mighty lion of the Ebstorf mappamundi! Fazio, transfixed and gawking, raises his hands in shock at the sight of her lowly condition.

The image would seem to visualize the cliché that reduced Rome to a moralizing lesson on the vicissitudes of fortune and the transience of worldly achievement. The thirteenth-century writer Magister Gregorius, quoted at the beginning of this chapter extolling the city's faded magnificence, also wrote that "all temporal things soon pass away, especially as Rome, the epitome of earthly glory, languishes and declines so much every day." The same mixture of pathos and wonder characterizes this image, for if one looks beyond the wretched old woman and dismayed onlookers, the city itself is a cornucopia of spruced up monuments. Alongside the most important pilgrimage churches, for example, the Colosseum appears fully intact. To the left of it, the equestrian statue of Marcus Aurelius is depicted below the Lateran Basilica, and above a stretch of aqueduct. Beneath that, the Pantheon rises prominently.

Ancient landmarks share center stage with Christian churches, betraying little sign of decay, or of the humility that characterizes *Roma* the woman. The

colosseo

Scm iohan laterani

Tempḷi paci

caphore

colona adma na

Sca su

colona anto nina

monte de caualli marmorei

castel sco agnola

Ponte salano

Ponte molle

tume .

orsi .

tanto

orsi .

nto

sta

anto .

a

egno

hiacta .

no

regno .

a noi

mei .

a

entile

n suoi .

tule

re

o stile .

e puo dire

image's sanctimonious façade is just that: a smokescreen for an enduring fascination with Roman marvels, pagan as well as Christian—in a departure from the exclusive focus on sacred topography we saw earlier. It was this dual identity that would prevail as Rome moved into the Renaissance.

Fig. 45

Opposite: View of Rome from Fazio degli Uberti, *Dittamondo*, 1447. Bibliothèque nationale de France, Paris.

FURTHER READING

Birch, Deborah J. *Pilgrimage to Rome in the Middle Ages: Continuity and Change*. Woodbridge, UK: Boydell Press, 1998.

Camerlenghi, Nicola. *St. Paul's Outside the Walls: A Roman Basilica, from Antiquity to the Modern Era*. Cambridge: Cambridge University Press, 2018.

Connolly, Daniel. *The Maps of Matthew Paris: Medieval Journeys through Space, Time and Liturgy*. Woodbridge, UK: Boydell Press, 2009.

Edson, Evelyn. *Mapping Time and Space: How Medieval Mapmakers Viewed Their World*. London: British Library, 1997.

Edson, Evelyn. "The Medieval World View: Contemplating the Mappamundi." *History Compass* 8/6 (2010): 503–17.

Gregorius, Magister. *The Marvels of Rome*. Translation, introduction, and commentary by John Osborne. Toronto: Pontifical Institute of Mediaeval Studies, 1987.

Krautheimer, Richard. *Rome: Profile of a City, 312–1308*. Princeton: Princeton University Press, 1980.

Krautheimer, Richard. *Three Christian Capitals: Topography and Politics*. Berkeley: University of California Press, 1983.

Kugler, Hartmut, Sonja Glauch, and Antje Willing, eds. *Die Ebstorfer Weltkarte: Kommentierte Neuausgabe in zwei Bänden*. 2 vols. Berlin: Akademie Verlag, 2007.

Kupfer, Marcia. "Medieval World Maps: Embedded Images, Interpretive Frames." *Word and Image* 10 (1994): 262–88.

Morse, Victoria. "The Role of Maps in Later Medieval Society: Twelfth to Fourteenth Century." In *The History of Cartography*, vol. 3, *Cartography in the European Renaissance*, ed. David Woodward, 25–52. Chicago: University of Chicago Press, 2007.

Nichols, Francis Morgan, and Eileen Gardiner. *The Marvels of Rome = Mirabilia Urbis Romae*. New York: Italica Press, 1986.

Wolf, Armin. "The Ebstorf Mappamundi and Gervase of Tilbury: The Controversy Revisited." *Imago Mundi* 64 (2012): 1–27.

Chapter Four

Rome Reborn

"Rome was all the world, and all the world is Rome."

Joachim du Bellay, *Antiquitez de Rome*, 1558

Although Renaissance means rebirth, suggesting something abrupt and earthshaking, this much vaunted period began not with a bang but a whimper, not in an instant but over a century, and not in Rome but in Florence. The Renaissance has been defined, redefined, and had its basic validity questioned by scholars in recent decades. The conventional notion, as set out by Swiss scholar Jacob Burckhardt in his *Civilization of the Renaissance in Italy* (1860), holds that it was marked by humanist scholarship and the rediscovery of classical antiquity—its art, literature, historical awareness, and emphasis on individual human potential—and that it represented a clean break with the Middle Ages.

While many of Burckhardt's ideas still hold up, scholars these days are more likely to speak of continuity than rupture and of debts to the preceding centuries: to see the Renaissance as a slowly unfolding cultural movement that grew out of—yet billed itself as the antithesis to—the so-called Dark Ages. They are also likely to point to larger demographic and social changes that marked the period as distinct and "early modern," such as increasing urbanization, new forms of popular piety, expanding mercantilism, and so on.

However you interpret the era, there is no doubt that one of its defining intellectual characteristics was a surging interest in the material as well as the literary remains of antiquity. So if Rome was something of a latecomer to the Renaissance, it was also its natural home. After the return of the popes, the city slowly but surely made up for lost time, surpassing Florence as epicenter by about 1500.

Cartography, in this period, experienced its own renaissance, due in part to major advances in surveying and in part to the print revolution—which was a cultural watershed on par with industrialization several centuries later. Maps of Rome discussed in this chapter embody the shift to print from manuscript, as well as new techniques for measuring the city and for representing its physical features. The view of Rome from the late fifteenth-century Nuremberg Chronicle is a landmark of early print and artistic verisimilitude, while Leonardo Bufalini's plan of the city is a milestone in urban cartography. The anonymous painting from Mantua and Antonio Tempesta's etched view both present panoramas that use new techniques of observation and representation to bring the city to life before the eyes of the beholder.

Collectively, all of these items bear witness to the evolution not just of cartography and art but also of the Church and the city, allowing us to trace a remarkable series of fast-paced urban and architectural changes to Rome's fabric. That said, these maps are much more than objective records, for they all frame the city through artistic selections, perspective, and viewpoint. In this way these images are also a case study in the fundamental expressiveness of cartography. As we shall see, their underlying messages oscillate between optimism for a city on the rise—in which the past will soon be equaled, perhaps even surpassed by the present and future—and longing for the city that once was.

The Renaissance in Rome did not begin auspiciously: the return of the popes to their traditional seat after an absence of more than a century was hardly triumphant. In 1420, Pope Martin V entered Rome together with the whole papal court, and so began the slow process of repairing the broken city. After having been left to its own devices for more than a century, Rome must have presented a sorry sight indeed. Its population was less than 20,000, its infrastructure ruined, its churches in disrepair. Clashing baronial families waged war from their fortified enclaves within the city. Much of Rome's territory inside the Aurelian Wall was a lawless wilderness where wolves and bandits roamed freely. Malaria ravaged

Fig. 46

Masolino da Panicale, *The Miraculous Founding of Santa Maria Maggiore*, 1423–25. Museo Nazionale di Capodimonte, Naples.

Rome Reborn

the surrounding countryside. It was an ungodly place, not for the faint of heart.

The popes, initially, proved faint of heart, struggling to gain a secure foothold. Even Martin V, a native Roman and member of the old and powerful Colonna family, dragged his feet for three years before coming to the city following his election at the Council of Constance, in southern Germany. His successor Eugenius IV, similarly fearing for his safety, spent the better part of his sixteen-year papacy in the relatively comfortable surroundings of Florence. Nevertheless, both men managed to reassert papal authority through symbolic means—by sponsoring works of art and architecture that projected the power of their office and the status of Rome as a site of Christian miracles and martyrdoms.

Martin turned his attention toward Rome's earliest churches, ordering improvements at St. John Lateran and the adjoining papal palace, as well as at St. Paul's Outside the Walls and St. Peter's. He also revived the Maestri delle Strade, or Masters of the Streets, a central agency charged with overseeing urban development and order. The same pope, or possibly a family member acting on his behalf, is believed to have commissioned the Florentine artist Masolino in the 1420s to paint a new altarpiece for Santa Maria Maggiore, one of the city's oldest and most prominent basilicas.

One side of the altarpiece showed the pope, beneath the protective gazes of Jesus and Mary, tracing the outline of the church's floor plan after a miraculous snowfall on August 5, 358—said to be a sign from the Virgin Mary (fig. 46). Not coincidentally, Santa Maria Maggiore lay in the heart of Colonna territory and functioned as a kind of family stronghold. Martin was typical of Renaissance patrons, in the sense that an act of Christian altruism, such as donating funds for church improvements, functioned simultaneously as an act of personal and familial commemoration.

Two decades later, Pope Eugenius IV commissioned the architect and sculptor Filarete to cast a huge new pair of bronze doors, adorned by rectangular relief panels, for St. Peter's Basilica. The central-right panel shows Eugenius kneeling before Peter and receiving the keys directly from the apostle: a message of authority by proxy that was especially important to convey in a time when the papacy in Rome was still on shaky ground (fig. 47).

The panel below shows Peter's crucifixion—upside down, at his own request, because he felt unworthy to die exactly as Christ had (fig. 48). While not topographically accurate, the scene is clearly set in Rome, for it incorporates bits of the city's famous ancient ruins. Beneath the condemning figure of Emperor Nero are depictions of the Castel Sant'Angelo along with the two pyramidal tombs then believed to be those of Romulus and Remus, all shown close to a river that can only be the Tiber. In this way—just as in Masolino's scene of the founding of one of Rome's most venerable churches—the city is shown to be a holy site, synonymous with the papacy and all it stands for. This message, too, was important to convey in order to put to rest once and for all the notion that the popes could call any other place home.

Both commissions raise an important issue about fifteenth-century Rome: talent had to be imported. The previous century—with its widespread, devastating episodes of plague and its political unrest—had not been a sparkling period for any central Italian city. But Rome had languished longer and deeper than most. Well into the 1400s it lagged behind the cultural momentum and artistic vibrancy of Florence. For that reason, Rome had to look north for the architects, painters, and sculptors who would bring new life and beauty to the city. The city, for its part, offered an increasing number of professional opportunities for the artisan classes, who accordingly settled there in ever greater numbers.

Nicholas V, pope from 1447–55, was the first to fully reassert papal authority, subdue the rival families, and make significant inroads when it came to

Rome Reborn

restoring Rome. He focused his efforts on fixing up the city's churches as well as improving its water supply and fortifications. He also managed to lure progressive intellectuals as well as craftsmen from Tuscany in sufficient numbers to ensure that the revival would take flight.

Nicholas's specific accomplishments were legion. He bolstered the authority of the Masters of the Streets, repaired the Acqua Vergine aqueduct, and designed—with his gifted architectural advisor Leon Battista Alberti—the first scheme for urban renewal, focusing on the Vatican Borgo (the neighborhood fronting St. Peter's). He assembled the nucleus of the Vatican Library and built a new Vatican Palace wing as well as a formidable round tower nearby to bolster the defenses of the ninth-century Leonine Wall. With the architect Bernardo Rossellino, Nicholas hatched a plan to shore up the ancient basilica of St. Peter's by strengthening its flimsy side walls and constructing an enormous new choir at the apse end of the church. Most of these grand plans barely made it off the drawing board, but their ambitious scale and scope set the tone for later papal initiatives.

Sixtus IV, pope from 1471–84, was the next to have a major impact. Like Nicholas, his focus was partially on Rome as a whole and partially on the Vatican. He built the papal showcase known, in his honor, as the Sistine Chapel between St. Peter's and the Vatican Palace, commissioning the most celebrated Florentine and Umbrian artists of the day—figures like Botticelli, Ghirlandaio, and Perugino—to adorn its side walls with glorious frescoes. He stimulated architectural patronage on the part of cardinals by enacting legislation permitting

Fig. 49

Ponte Sisto, Rome. Photo: Livioandronico2013 / Wikimedia Commons. CC BY-SA 4.0: https:// creativecommons.org/ licenses/by-sa/4.0/legalcode.

Chapter Four

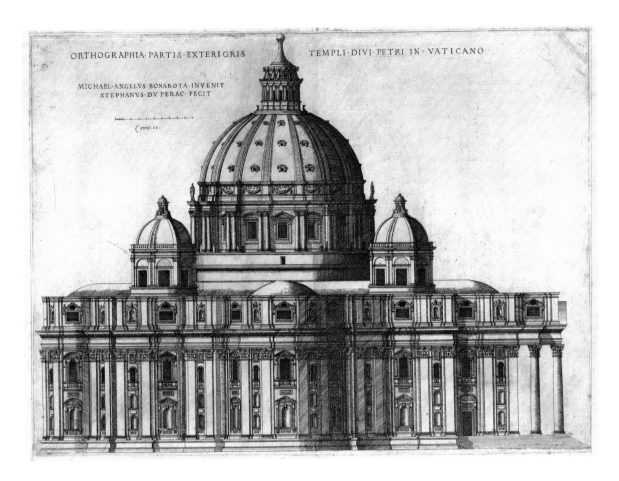

ORTHOGRAPHIA·PARTIS·EXTERIORIS   TEMPLI·DIVI·PETRI·IN·VATICANO

MICHAEL·ANGELVS·BONAROTA·INVENIT
STEPHANVS·DV PERAC·FECIT

Fig. 50

Stefano Du Pérac,
Michelangelo's design for
St. Peter's, engraving with
etching, 1569. Metropolitan
Museum of Art, Rogers
Fund, Transferred from
the Library, 1941.

them to bequeath property to their heirs, rather than having it revert back to the Church after their death. A boom in palace building followed suit.

Like almost all papal projects, those of Sixtus were geared toward both Rome's permanent population and its pilgrims. He granted the Masters of the Streets power to seize and destroy private property for the sake of the public good and to make way for new churches or civic constructions; brought to the Capitoline Hill a number of ancient statues that were scattered around the city, thereby establishing the first public sculpture collection, still there today; and built a new bridge (on surviving ancient foundations) connecting Trastevere and Rome's center—also still there.

That bridge, known as the Ponte Sisto—again in honor of the pope—provided a convenient pathway across the Tiber that greatly benefitted the *Trasteverini* on an ongoing basis (fig. 49). It also served the needs of visitors who thronged the city during Jubilees, for it opened an alternate route toward St. Peter's, easing circulation. Sixtus was keenly aware of this necessity: during the Jubilee of 1450, more than a hundred pilgrims had been crushed to death when a stampede broke out on the heavily overcrowded Ponte Sant'Angelo—the main conduit leading from the city center to the Borgo. The Ponte Sisto opened downriver just in time to avoid the danger of a repeat tragedy during the Jubilee of 1475.

Rome Reborn

The most visible and controversial project of Renaissance Rome was New St. Peter's basilica. Its story began in 1506, when Pope Julius II abandoned the cautious approach of his predecessors, opting not to renovate but rather to raze and rebuild the venerated ancient structure. Construction and demolition proceeded in fits and starts and involved many of the most celebrated architects of the time, from Bramante to Michelangelo and Bernini (fig. 50). For now, suffice it to say that this massive, century-long project encapsulates many of the contradictions of the time. The whole concept was based on a mixture of hubris and piety. Its architectural style reflected reverence for antiquity, its substance—largely building materials pillaged from ruins—blithe disregard for the past. The immense church was meant to serve as a beacon for the faithful, even as its extravagance helped spark Martin Luther's fateful critique of the papacy: a critique that would launch the Protestant Reformation and violently split the Western Christian Church.

These paradoxes notwithstanding, New St. Peter's lurched forward, much like Rome itself for most of the sixteenth century—despite such catastrophes as the Sack of 1527, when Charles V's marauding troops brought the city briefly to its knees, and the more long-term existential threat posed by the Reformation.

Fig. 51

View of Rome from Hartmann Schedel, *Liber chronicarum*, woodcut, Nuremberg, 1493. The Barry Lawrence Ruderman Map Collection, David Rumsey Map Collection, www.davidrumsey.com.

Chapter Four

Ultimately, St. Peter's emerged from the Renaissance completely reinvented, just like Rome itself.

A City Ready for Its Close-Up

Based on a now-lost engraving of Rome published by Francesco Rosselli in Florence about 1485–90, this view (fig. 51) was included in Hartmann Schedel's *Weltchronik* (or *World Chronicle*), a landmark volume in the early history of print. Published in 1493, just four decades after Johannes Gutenberg printed his celebrated Bible on his newly invented press, the lavish tome known familiarly as the *Nuremberg Chronicle* was a Renaissance bestseller that recounted the history of the world through a Christian lens, from Creation through the present. Its hundreds of illustrations included numerous city views, many of them the first recognizable images of the places depicted.

The view of Rome was printed as a double-page spread and presents a picture of a bustling, scenic town as it might appear to a visitor approaching from the Pincian Hill on the northeast side of the city. The Aurelian Wall extends

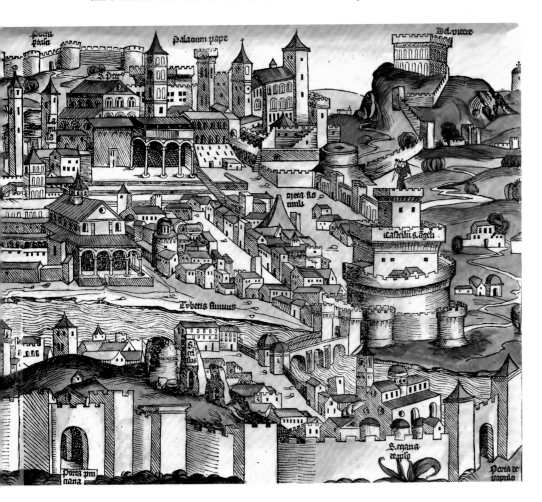

St. Peter's

Sistine Chapel

Vatican Palace

Belvedere Villa

Benediction Loggia

Tower of Nicholas V

Fig. 52

View of Rome from Schedel, *Liber chronicarum*, detail showing the Vatican Hill

across the foreground, its open gates beckoning the viewer to enter. Beyond that porous barrier, famous monuments pepper the cityscape, nestled together cozily in the undulating topography. At far left (south) is an edge of the Colosseum, at far right (north) the Castel Sant'Angelo. Indeed, the view provides an abbreviated "greatest hits" of Rome, with celebrated ancient statues and ruins popping up here and there. The well-known statue of the horse tamers on the Quirinal appears at lower left, while the Pantheon and the Column of Marcus Aurelius are prominently featured toward center. All of their proportions are inflated, as if glimpsed through a tourist telescope.

The Tiber runs diagonally across the page, the Vatican Hill rising beyond it, with St. Peter's and its bell tower crowning the image (fig. 52). The basilica's prominence—made possible, in part, by the view's low vantage point from the northeast—firmly establishes this city as a Christian capital. The woodcut gives a good overview of the structures dating from the fourth to fifteenth centuries that stood nearby. The oldest present was, of course, St. Peter's itself. In the image, its façade rises above a hodgepodge of later appendages. Plainly visible are the triangular roof above its central nave and one of its two lower, adjoining side wings, corresponding to interior double-aisles. The church is treated somewhat summarily, however: there is no indication, for example, of its transept (or cross arm).

In front of the basilica was a rectangular forecourt, or atrium, also dating from the early Christian period. Here, that open space is masked by the new

Benediction Loggia begun for Pope Pius II in 1462—one of the first major Renaissance improvements to the church complex. From this balcony the pope appeared to the faithful at key moments, such as when he was newly elected or when it was time to bestow the Easter blessing.

In the woodcut, that recent addition is just one of many that can be spotted in and around the Vatican, which the popes had made their official seat after returning from Avignon. To the right of St. Peter's and partly blocked by its tall medieval bell tower is the heavily fortified Sistine Chapel, completed in 1480. To the right of that structure, in turn, is the papal palace wing built for Nicholas V in the early 1450s. Just below and to the right of that cluster of buildings, Nicholas's imposing round tower can be glimpsed protruding from the Leonine Wall. Further to the right—just squeezed in at the upper corner—is the Belvedere Villa, isolated on a hill. Constructed for Pope Innocent VIII in the 1480s, this suburban papal retreat is the latest building represented. Monuments new and old intermingle here, creating a sense of a city on the upswing, buzzing with activity.

The City Seen through a Wide-Angle Lens

Dating to about 1540, this large painting of Rome (fig. 53) looks very different from Schedel's view, but it was actually based on the same influential, lost model by Francesco Rosselli. Note, for example, similarities in the viewer's positioning with regard to the Aurelian Wall and Pincian Hill, as well as the depiction of the Pantheon, toward center-right. Yet it conveys a very different, less idealized picture of the state of the city.

The painting, by an artist whose identity has not been recorded, was part of a mural cycle of illustrious cities commissioned by the Marchesa Isabella d'Este for Mantua's ducal palace. Rather than showing Rome from a low viewpoint just outside city gates, the painter opted for an omniscient bird's-eye perspective from far above, one that encompasses the whole circuit of the Aurelian Wall. This complete picture makes Rome appear less, not more, impressive, because it reveals the full extent of the sparsely settled greenbelt that sprawled over the city's eastern and southern zones (the left half of fig. 53). Gradually abandoned since late antiquity, this vast area was a potent signal of Rome's decline. The painting presents it as a swath of pastureland and wilderness dotted by a few isolated but important churches like St. John Lateran and Santa Maria Maggiore, as well as bits of ruins and broken aqueducts: symbols of lost luster.

Glimpsed in this larger urban context, Rome's center is revealed to be an island of habitation amid an encroaching sea of desolation. To be sure, it is an impressive island. Again, familiar sights like the Pantheon and the Column of Marcus Aurelius are clearly identifiable and disproportionately large (fig. 54). At the same time, the painting is more detailed than the smaller woodcut, granting a more thorough record of recent architectural and urban changes. Below

and to the left of the Pantheon, for example, is a monument missing from the woodcut: the massive Palazzo Venezia, begun by Cardinal Pietro Barbo and enlarged after he became Pope Paul II in 1464. That and other Renaissance constructions jostle for space with medieval baronial towers, ancient columns, and nameless structures in Rome's crowded fabric.

The painting's mixture of urban decay and revival hints at a certain ambivalence about Rome's current state. It is certainly a very different message than the woodcut's rosy picture. With their distinct outlooks, these two images together hint at the spectrum of possible responses to Rome's fortunes in the fifteenth and early sixteenth centuries. Was the city renewed, or was it a relic? The verdict was still out.

The City Measured

So much of Rome's Renaissance history comes together in Leonardo Bufalini's large and magnificent woodcut map (fig. 55) that one scarcely knows where to begin. So let's begin with a catastrophe. The map was published at midcentury,

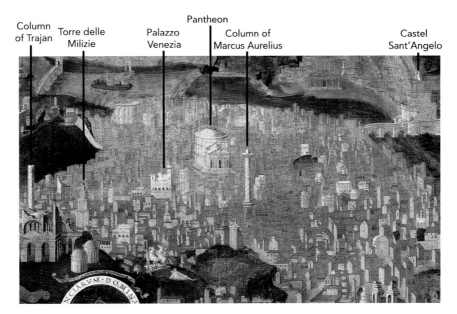

Column of Trajan — Torre delle Milizie — Palazzo Venezia — Pantheon — Column of Marcus Aurelius — Castel Sant'Angelo

Fig. 54

Anonymous, view of Rome, detail showing the city center

almost twenty-five years after Rome's most brutal setback: the 1527 Sack. On May 6 of that year, the imperial soldiers of Charles V began a vicious assault on the city, ignoring the wishes of their commanders and proceeding to lay waste to Rome and its inhabitants. Religious sentiment was secondary to violent mob instincts, but many of the soldiers were Lutheran mercenaries from Germany (despite being in the employ of the Holy Roman Emperor).

For them, the riotous circumstances became an opportunity to debase anyone and anything related to the papacy and the Catholic faith—a pretext for raping nuns and killing priests, defiling holy relics, smashing works of sacred art, looting precious metalwork, staging mock masses and conclaves, and on and on. Their message, if they had one, was that Rome and its corrupt institutions were dead.

The most acute period of the Sack lasted about a week, but the invading troops lingered in the city for the better part of a year, and their traces lasted considerably longer. Their graffiti can still be found, for example, scratched into the lower reaches of the Stanza della Segnatura, Raphael's magnum opus in the Vatican Palace, and Baldassare Peruzzi's illusionistic frescoes in the Villa Farnesina, a mile away near the banks of the Tiber.

The Sack was profoundly embarrassing to Charles V and Pope Clement VII alike, but the most lasting damage was not to their reputations—it was, rather, to Rome itself. The city's physical devastation was horrific, with an estimated 30,000 residences destroyed. "Hell itself was a more beautiful sight to behold," observed the Venetian Marin Sanudo. The city's population also took a major hit. It had risen to 55,000 on the eve of the Sack, and 6,000–10,000 are estimated to have been murdered in the first few days alone. Another 15,000 are

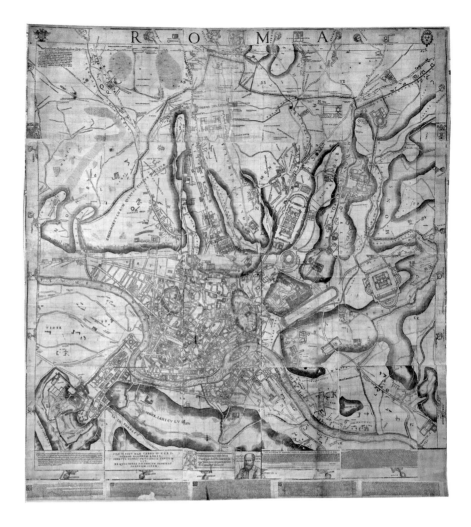

thought to have fled in the aftermath. This means, in part, that much of the
talent that had converged on Rome over the previous decades—all that intellectual and artistic capital, all that cultural momentum—was drained away in
what amounted to the blink of an eye.

The Sack of 1527 is often cited as the end of the Renaissance in Rome—or
even as the end of the Renaissance in general. To this day, many college courses
on Renaissance art end with the Sack. But rarely does an entire epoch conclude
so suddenly, and scholars continue to debate the extent to which the Sack was
a turning point. Much as the Renaissance was slow to get rolling, it was slow
to draw to a close—and when it did, it was not like a curtain coming down, but
rather a slow transition to something new. In any case, the repercussions of the
Sack in Rome itself seem to have been surprisingly short-lived.

Ample evidence of rapid urban recovery comes from Bufalini's 1551 map,
which takes the form of a ground plan, or footprint, of the city, oriented with
north at upper left (fig. 55). This richly detailed map records prestigious architectural projects of the time—including private palaces, like Palazzo Massimo
alle Colonne, begun just a few years after the Sack, and the Palazzo Farnese,

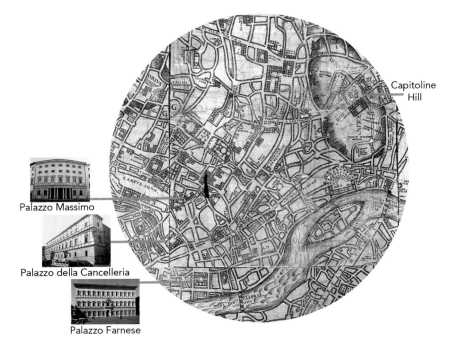

Capitoline
Hill

Palazzo Massimo

Palazzo della Cancelleria

Palazzo Farnese

Fig. 56

Above: Bufalini, map of
Rome, detail showing the
center

Fig. 57

Below: Bufalini, map of
Rome, detail showing the
Vatican with St. Peter's
circled

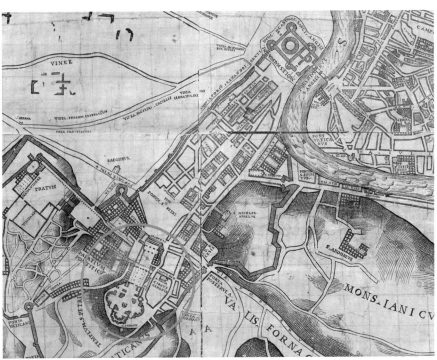

begun in the 1510s but expanded dramatically beginning in 1534, when its pro-
prietor Cardinal Alessandro Farnese became Pope Paul III (fig. 56). Mere steps
away is the slightly earlier Palazzo della Cancelleria, begun in 1489 for Cardinal
Raffaele Riario. The map also records the beginnings of Michelangelo's radical
makeover of the Capitoline Hill, discussed further below, which began in the

Rome Reborn

Fig. 58

Above: Bufalini, map of
Rome, detail showing the
street network with new
streets highlighted

Fig. 59

Below: Bufalini, map of
Rome, detail showing the
greenbelt with real and
imagined monuments

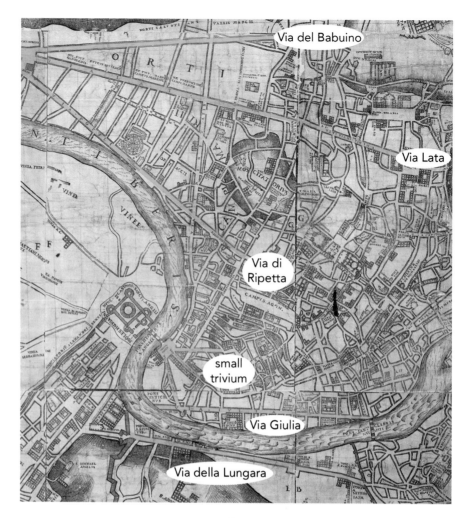

1530s. Rome's construction boom, it seems, paused just momentarily in the wake of the calamity.

Bufalini's map also gives a glimpse of Rome's other great work in progress: New St. Peter's (fig. 57; see also fig. 50). This structure had indeed suffered a building hiatus, though one that had commenced well before the Sack. Things started to move forward again after Michelangelo was appointed chief architect in 1546. Following the original design by Bramante, his called for a Greek-cross plan: a central crossing supporting a massive dome, from which four equal arms extended, as opposed to a more traditional Latin-cross plan, with a longer central nave. The final church was a compromise solution, but Bufalini accurately shows the midcentury plan. He also documents the nave of Old St. Peter's still standing—capturing the strange scenario of a doomed building temporarily coexisting with its own replacement.

Bufalini's cartographic approach to the urban fabric grants Rome's street plan a high degree of legibility. By the mid-1500s, many popes had worked to modernize this network. They faced a considerable challenge: Rome was an unplanned city from its origins, and its streets had grown organically over time. The ancient Via Flaminia, or Via Lata as it was known within the walls, was an exception, forming a long straight spine from the Porta del Popolo at the northern end of the city, south to the base of the Capitoline. By the late Middle Ages, however, that axis had been supplanted as the city's main thoroughfare by Via Papalis, a long, narrow, meandering street leading from the Lateran Basilica at the southeast to the Vatican and St. Peter's at the northwest. Linking Rome's two holiest sites, this was the path followed by many papal processions, most notably the *Possesso*, when a newly elected pontiff traced his own Christian triumphal route through the city to take symbolic possession of it.

Renaissance urban interventions had to uproot longstanding patterns in order to graft planning onto the existing fabric. Papal efforts for the fifteenth and most of the sixteenth century tended to be piecemeal, and to aggregate over time. They are visible throughout Bufalini's map (fig. 58). Toward the bottom of the image, two new straight streets are shown flanking both sides of the Tiber: Via Giulia on the left bank, and Via della Lungara on the right. The most conspicuous improvement is the three long, straight thoroughfares radiating out into the city from its northernmost point at Piazza del Popolo (visible at center-left).

The centermost one is Via Lata—today known as Via del Corso, Rome's main shopping street—which had been widened and regularized by Paul II in the 1460s and 1470s. Here, it is flanked by the new avenues of Via del Babuino to the east and Via di Ripetta to the west. The latter leads toward another, smaller "trivium" (Latin for "three streets") at the foot of the Ponte Sant'Angelo—essentially funneling foot

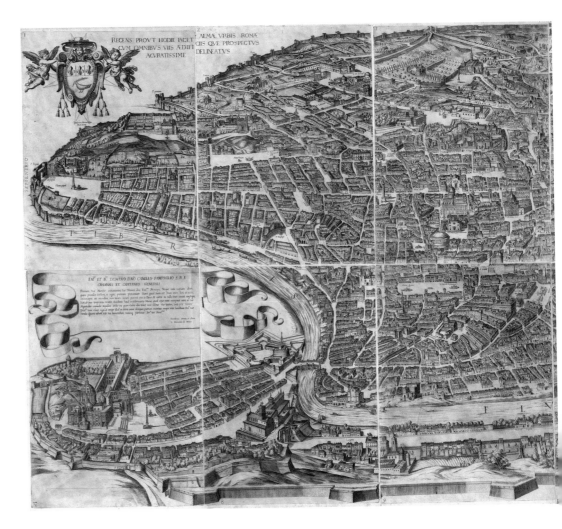

traffic toward and away from that main crossing to the Vatican. Slowly, these networks were granting Rome an improved circulatory system to benefit locals and visitors alike, while bringing the city into the modern age (or at least the early modern one).

More remarkable than anything represented on the map, however, is its very mode of representation. Today, in the era of Google Maps, city plans are ubiquitous, but in Bufalini's time they were a relatively new form, restricted mainly to expert circles of architects and military engineers, as opposed to laymen. Bufalini's measured plan with all elements shown to scale was a revolutionary departure from the legion of pictorial and bird's-eye views that were destined for public consumption.

That does not mean, however, that it was an entirely objective picture of sixteenth-century Rome. While Bufalini's plan does accurately and faithfully record many transformations to the Renaissance city, it also includes a number of fictitious, sometimes extravagant buildings that had little grounding in physical evidence (fig. 59). The blank space of Rome's greenbelt, in particular, seems to have sparked Bufalini's creative impulses. The largest structure depicted,

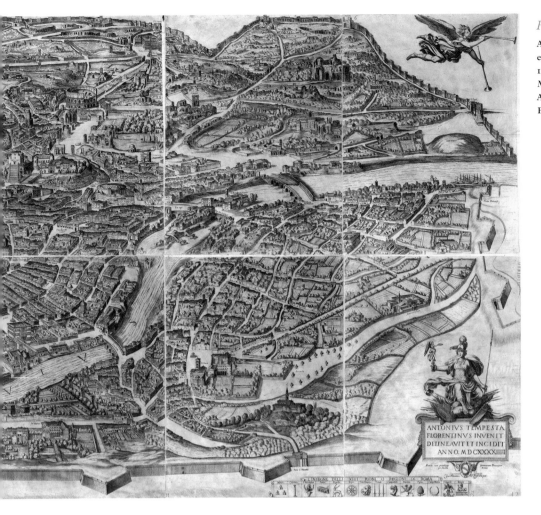

Fig. 61

Antonio Tempesta, bird's-
eye view of Rome, etching,
1593 (edition of 1645).
Metropolitan Museum of
Art, Edward Pearce Casey
Fund, 1983.

the Baths of Trajan, was in reality little more than a few fragmentary ruins, but
Bufalini depicts it as an enormous, complicated, complete structure. His map,
which initially appears so scientific, proves that nostalgia for a vanished past was
never far beneath the surface in Renaissance culture. It also serves as a reminder that a map, far from objective, is often the brainchild of a single, subjective
creator: here, an intellectual who proudly includes his own self-portrait at the
bottom center of his map (fig. 60), in an extraordinary claim of authorship.

A Panoramic View of Urban Revitalization

By 1600, any ambiguity about Rome's status had been resolved, and the city's
renewal was in full swing. Its population had not only recovered from the Sack
but almost doubled to about 100,000. New churches, palaces, and suburban
villas were springing up around the city at a dizzying pace. Antonio Tempesta's

Rome Reborn

grand panoramic view captures the excitement of this time—the sense of having emerged triumphant from grave challenges (fig. 61). Here, too, there is no trace of ambivalence, only optimism. Rome in all its glory unfolds beneath the eyes of the viewer from a vantage point that corresponds to the Janiculum Hill, high above Trastevere on the western side of the city: still the best spot from which to enjoy a scenic vista. Even in ancient times, the poet Martial had marveled, "From here you can see the seven lordly hills and measure the whole of Rome."

Bufalini had been the first to adopt this vantage point in his map, although Tempesta was the first to exploit its evocative potential. In fact, the spectator seems to take flight and ascend well above the Janiculum's crest, seeing the city more fully than was possible from even the highest hill. It looks to modern eyes like the view from an airplane, or perhaps a drone, but of course such a vision was completely imaginary—and nothing short of astonishing—in this era before flight.

One place that comes into clear focus toward the center of the image is the foremost urban renewal project of the sixteenth century: Michelangelo's renovation of the Capitoline Hill, commonly known as the Campidoglio. This is the smallest of Rome's seven hills but also the densest with symbolism. In antiquity it was the city's religious center, in the Middle Ages its civic center. By the 1500s, it was run down (fig. 62). Several medieval and early Renaissance structures dominated the overgrown hilltop. The Franciscan church and monastery of Santa Maria in Aracoeli sat at the highest point on the north, accessed from the

Fig. 62

Anonymous, view of the Capitoline Hill, early sixteenth century. Photo: RMN-Grand Palais / Art Resource, NY.

Chapter Four

Fig. 63

Stefano du Pérac, Michel-
angelo's design for the
Campidoglio, etching, 1569.
Metropolitan Museum of Art,
Rogers Fund, Transferred
from the Library, 1941.

base of the hill via a steep, towering staircase. Beneath its flank, to the east, was
the Senatorial Palace (Palazzo Senatorio), partly refurbished by Nicholas V in
the mid-1400s, capped by a tower and fronted by an unkempt, unpaved piazza.
To the south, the same pope had built the Palace of the Conservators (Palazzo
dei Conservatori), home to an elected judicial body and subsequently to the
sculpture collection established by Sixtus IV.

A turning point for this site came in 1536, when Paul III granted Charles
V, fresh from a victory over the Ottomans at Tunis, the honor of a procession
through Rome in the manner of an ancient emperor returning triumphant from
a military campaign. The gesture was intended to signal reconciliation between
the two powers following the previous decade's Sack. In antiquity such processions culminated at the Capitoline, but it was in no state to serve that purpose
this time around. Realizing it was sorely in need of a face-lift, the pope commissioned Michelangelo to transform the Campidoglio into a fittingly grand
public space.

The resulting design for a symmetrical, orderly oasis meant to resonate with
the geometry of the universe still defines the Campidoglio today (fig. 63). Visitors approaching from the city climb a grand staircase, at the top of which they
are greeted by two "gatekeepers"—versions of the horse-tamer statues—and
by the famous bronze equestrian statue of Emperor Marcus Aurelius, gazing
out calmly and commandingly, at the center of the gracious, trapezoidal piazza,
which rises ever so slightly toward that midpoint.

At the far end of the space, the Palazzo Senatorio—which Michelangelo clad
in a new, Renaissance façade, with important statues grouped at the base—
creates a theatrical backdrop to the emperor. The Palazzo dei Conservatori to
the right also has an updated façade in the classical mode, which is mirrored
by a new palace at left, the Palazzo Nuovo. Little more than a shell, the second

Rome Reborn

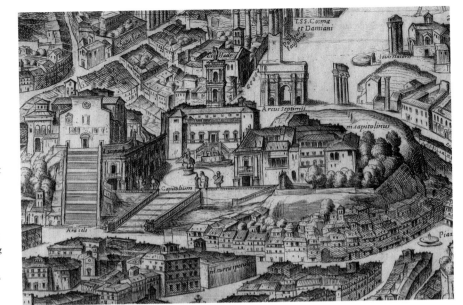

Fig. 64

Above: Tempesta, bird's-eye view of Rome, detail showing the Capitoline Hill

Fig. 65

Below: Tempesta, bird's-eye view of Rome, detail showing Santa Maria Maggiore (upper left), St. John Lateran (upper right), and the new streets of Sixtus V

palace exists just to create symmetry. Overall, Michelangelo's design is a monumental stage set and a glorious pinnacle of Renaissance urbanism.

Tempesta's view, specifically the 1645 version reproduced here, shows the project largely complete a century after it was conceived (fig. 64). The grand new staircase leads up to the symmetrical Palazzo Senatorio and the recently completed Palazzo Nuovo, facing the Palazzo dei Conservatori from the shadows of the adjacent Santa Maria in Aracoeli. One element that would not be completed for centuries was Michelangelo's radiating pavement design, with its cosmic undertones reminiscent of celestial orbits.

Tempesta's map also bears witness to a formative moment in Rome's urban history. In the late 1580s, Pope Sixtus V initiated a series of sweeping projects that set the city on a course toward dramatic growth. What set his agenda apart

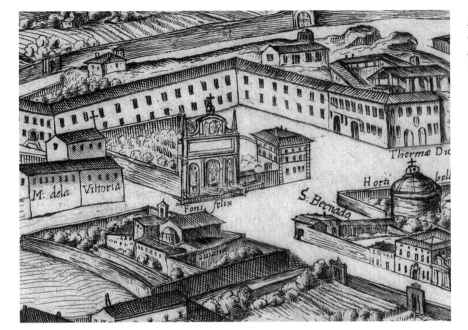

Fig. 66

Tempesta, bird's-eye view
of Rome, detail showing the
Fountain of Moses

from those of his predecessors was that he moved beyond repairing the inhabited parts of Rome, focusing instead on the city's depopulated greenbelt as a site of future development. It was a visionary shift from retrofitting to planning: from looking back to looking ahead.

In the upper half of Tempesta's map, a new web of long, straight streets slashes through the landscape. Laid out by Sixtus's chief architect, Domenico Fontana, they converge on Santa Maria Maggiore and its surrounding settlement, which previously had been somewhat isolated atop the bucolic Esquiline Hill (fig. 65). From there they connect to other important basilicas, like St. John Lateran, as well as to landmarks like the Colosseum. Many of these focal points in the Sistine network are marked by obelisks. Pillaged from Egypt in antiquity, they had been placed symbolically around Rome as monumental trophies. During the Middle Ages, most had toppled. Their re-erection by Sixtus V was in itself a triumph.

Sixtus's streets served the double purpose of providing clear and efficient routes for pilgrims and paving the way for later permanent settlement (they were not, however, paved: for the most part Roman streets remained unpaved well into the nineteenth century, with most of the city's characteristic *sampietrini*, or cobblestones, being a relatively recent addition). The new streets also reflect a recent development: the carriage traffic that increasingly choked Rome's narrow, tortuous network. The impact of Sixtus's scheme far transcended Rome itself. His arterial system of long, straight streets meeting at important, visually prominent nodes influenced later urban plans from L'Enfant's Washington, DC, to Haussmann's Paris.

In another crucial precondition for expansion in Rome's greenbelt, Sixtus restored the ancient aqueduct known as the Acqua Alessandrina in order to supply water to this part of the city. It was rechristened, in his honor, the Acqua

Felice (the pope's given name was Felice Peretti, and *felice* also means "happy" or "joyful"). Its presence is suggested on Tempesta's map by the inclusion of the grand new fountain designed by Fontana on the Quirinal, which fittingly shows Moses bringing water to his people (fig. 66)—Sixtus, by implication, being the new Moses to the Romans.

Tempesta's view also shows the ongoing evolution of Rome's center. One noteworthy development was the cluster of churches that had recently cropped up, most of which pertained to the new religious orders that had come to prominence during the militant period of the Counter-Reformation, Rome's official response to the Protestant threat. They include the Gesù, mother church of the Jesuit order, as well as Sant'Andrea della Valle and Santa Maria in Vallicella (or Chiesa Nuova), flagships of the Theatines and Oratorians, respectively (fig. 67).

Of course, the most important church in Rome was still St. Peter's (fig. 68). Tempesta portrays the new building complete at last and crowned by Michelangelo's dome: a feat of construction that Sixtus had finally managed to push through twenty-five years after the architect's death. In front of the basilica is

Fig. 67

Tempesta, bird's-eye view of Rome, detail showing the Gesù (top right), Sant'Andrea della Valle (center), and Chiesa Nuova (bottom left)

Chapter Four

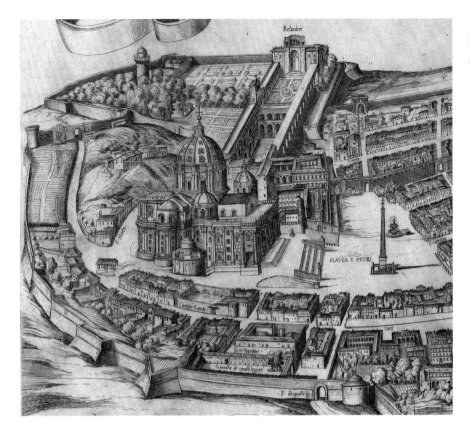

Fig. 68

Tempesta, bird's-eye view
of Rome, detail showing
St. Peter's

the obelisk that Fontana had relocated to that spot with great fanfare, but otherwise the open space in front of the church appears irregular and unfinished—Bernini's great keyhole-shaped piazza was still in the future. On the far side of St. Peter's, by contrast, the hill ascending to the Belvedere Villa has been tamed. There, two long, parallel galleries flanking a series of terraced courts—all designed by Donato Bramante in the first decade of the sixteenth century—have been bisected by a new Vatican Library wing. The whole complex has now assumed a form very close to its modern layout as the Vatican Museums, as well as the Library and the Apostolic Archive.

Tempesta's picture is seductive, but it smooths over many gritty realities of Renaissance Rome. With maps, it is always important to consider what is omitted. There is no vestige of the short-lived *Ortaccio*, or prostitutes' quarter, established by Pius V in the 1560s in an attempt to contain and control the legions of female sex workers who serviced Rome's majority male population. The intrepid viewer can find signs of the Jewish ghetto, a gated precinct located just across the river from Tiber Island, which had been established by Paul IV in 1555 to similarly oppress another, more venerable, equally longstanding Roman community. Tempesta simply and unobtrusively depicts its gates, giving no hint of the squalid conditions that prevailed in this prison.

Maps are always smokescreens in one way or another. Perhaps it would have been hard to map something as intangible as the repression of ideas that came with the Inquisition, when luminaries like Giordano Bruno and Galileo

Rome Reborn

Galilei were tortured and sometimes killed for their revolutionary theories. Overall, squalor and persecution had no place in a commemorative image like Tempesta's—but they were as much a part of life in Renaissance Rome as the grand thoroughfares, glittering churches, fountains, and palaces.

FURTHER READING

Burckhardt, Jacob. *The Civilization of the Renaissance in Italy.* Trans. S. G. C. Middlemore. London: Penguin Books, 1990.

Chastel, Andre. *The Sack of Rome.* Princeton, Princeton University Press, 1983.

Cohen, Elizabeth S. "Seen and Known: Prostitutes in the Cityscape of Late-Sixteenth-Century Rome." *Renaissance Studies* 12 (1998): 392–409.

Hook, Judith. *The Sack of Rome.* 2nd edition. New York: Palgrave Macmillan, 2004.

Karmon, David. *The Ruin of the Eternal City: Antiquity and Preservation in Renaissance Rome.* New York: Oxford University Press, 2011.

Maier, Jessica. *Rome Measured and Imagined: Early Modern Maps of the Eternal City.* Chicago: University of Chicago Press, 2015.

Partner, Peter. *Renaissance Rome, 1500–1559: A Portrait of a Society.* Berkeley: University of California Press, 1976.

Redig de Campos, Deoclecio, Maria Donati Barcellona, et al. *Art Treasures of the Vatican: Architecture, Painting, Sculpture.* New York: Park Lane, 1981.

Rowland, Ingrid. "Rome at the Center of a Civilization." In *The Renaissance World*, ed. John Jeffries Martin, 31–50. New York: Routledge, 2009.

Chapter Five

―――――

Rome of the Scholars

"Fate weighs down on all things created, Rome, you teach how time devours all things. Rome, you teach how once you were renowned for your high roofs, and now you lie underneath a great pile of ruins."

Cristoforo Landino (1424–98)

"How great Rome was, the ruins themselves show."

Sebastiano Serlio, 1540

The scholar Cristoforo Landino and the architect Sebastiano Serlio used different words to express the same feeling: a keen awareness that Rome had once been much grander. It is something of a paradox that at the very moment when Rome was gaining a new lease on life as capital of the High Renaissance, many intellectuals were starting to look back wistfully at the bygone glories of the ancient city. In fact, these two conditions—modernity and sentimentality—grew hand in hand, two sides of the same coin. And given that this period was marked by a dual fixation with its own place in history and with antiquity, it is hardly surprising that nostalgia took hold like never before.

Ruins were the wellspring of that feeling: an inescapable presence, looming over every corner of the city (fig. 69). Even in their decaying state, they were Rome's most impressive sites—rivaled only recently by New St. Peter's and

Vestigij d'una parte di dentro delle terme d'Antonino caracala qual fu adornata di grandissime et belle colonne di granito orientalle con le sue menbri intagliati con bella diligentia
et li muri furono incurs. ti di diverse pietre di mischi et marmori come hoggi sene vede anch[o] vestigij, et non molti annj sono fu donato da Popa Pio IIII una di
dette colonne al gran Duca quale fu da lui mandate in Fiorenza Il luoco doue erauno dette colonne si vede a questo segno A.

19.

Fig. 69

Aegidius Sadeler II after
Stefano Du Pérac, ruins
of the Baths of Caracalla,
etching, 1606. Metropolitan
Museum of Art, Bequest of
Phyllis Massar, 2011.

other ambitious Renaissance constructions. The fragmentary appearance of
ruins made them all the more alluring. The missing parts invited observers to
fill in the blanks mentally, to imagine those crumbling buildings back to wholeness. Almost inevitably, the result was a romanticized picture. After all, ruins
conveyed a mute poetry. They seemed to make a silent, moralizing commentary
on the present, on the passage of time, and on hubris of all kinds. People could
project onto them any number of meanings and messages.

As we shall see in this chapter, mapmakers in the sixteenth century and beyond did just that, creating deeply personal and unique responses to the visible
remains of ancient Rome. For Bartolomeo Marliani, the ruins were there to be
documented as incontrovertible physical facts. For his archrival Pirro Ligorio,
those physical facts—those impoverished relics—were just a starting point to
revive the glories of the ancient city. Ligorio, in turn, was the forefather of perhaps the greatest interpreter of Rome's ruins, Giovanni Battista Piranesi, who
two centuries later left behind any sense of obligation to the purported reality of
history. For him, the ruins existed primarily to spark his own fertile imagination
and to help him create a map where the language of cartography expressed a
vision of the past as futuristic clockwork. All of these men were highly learned,
and their stark differences reveal the extent to which creativity was once integral
to scholarship—not at odds with it.

The Renaissance fascination with ruins was just one part of the larger longing to revive Rome's past, but it was an extremely important part. The desire to know more about the ruins motivated artists and architects to do what no one had really thought of doing before: investigate their structural and stylistic properties through personal observation, even excavation. In the early fifteenth century, the architect Filippo Brunelleschi, according to his biographer Antonio Manetti, sought "to rediscover the excellent and highly ingenious building methods of the ancients and their harmonious proportions." To that end, Brunelleschi "made rough drawings of almost all the buildings in Rome," and he "had to dig in many places in order to investigate structures." None of this activity would seem all that noteworthy today, but six hundred years ago, it was unheard of to the extent that it aroused suspicion.

Assuming Manetti's fable to be true, then Brunelleschi was among the earliest of many artisans who pioneered basic principles of modern archaeology, literally getting their hands dirty to gather data empirically. Their ultimate goal was not necessarily to complete the historical record or—like archaeologists today—to better understand vanished cultures. Their interests were more personal and pragmatic: they were seeking intimate knowledge of ancient art and architecture so they could imitate it in their own drawings and designs. This explains why the first archaeologists were not classically trained scholars, who gravitated more toward texts, coins, and inscriptions that they could examine in the quiet of their studies.

A century after Brunelleschi, painters including Pinturicchio, Ghirlandaio, and Raphael eagerly awaited their turn to be lowered through a hole in the ground into the dark, cavernous, subterranean chambers of Nero's Golden House, where they could stare in wonder—by candlelight—at the lavish surroundings and sketch the remarkably well preserved stucco and fresco decorations from the first century CE. Speaking from personal experience, an anonymous painter from Milan wrote, about 1500: "No heart is so hard it would not weep for the vast palaces and broken walls of Rome, triumphant when she ruled; now they are caves, destroyed grottoes with stucco in relief." Artists like that one sought to revive those vast palaces and broken walls by means of their own paintings, architectural settings, and designs—with perhaps the most famous example being Raphael's Villa Madama, a grandiose and only partially realized scheme to rival antiquity (fig. 70).

The obsession with ruins also deepened the roots of Western classicism—that is, the desire to emulate stylistic aspects of Greco-Roman antiquity in literature and the arts, and thereby to embody ideals associated with those periods (aesthetic, political, and otherwise). It is a common and pervasive misconception that the classical tradition died out in medieval Europe only to be rediscovered in the Renaissance. In truth, there were many upswells of classicism in the Middle Ages, when it was one among a variety of stylistic modes available to artists and patrons. Still, it gained a degree of traction and preeminence in the Renaissance that was new, achieving the full force of a cultural movement.

From that time, classicism has never really receded from view, retaining its prestige over the centuries: from Palladian villas in seventeenth- and eighteenth-century England, to neoclassical painting in France in the late eighteenth and nineteenth centuries, Beaux Arts architecture in the nineteenth-century United States, and "new classical" architecture in the twenty-first century. At the same time, a movement's influence is partly to be measured in reactions again it. Classicism is a stylistic chameleon that can be used for good or for ill. It has been equated with liberty or colonialism, pressed into service as a symbolic language of democracy or of oppression—its potency mobilized by people, institutions, and regimes seeking to convey stability, authority, and venerability.

As a result, classicism has incited many artistic revolutions aimed at overthrowing its cultural dominance: from realism and impressionism to modernism and postmodernism, conceptual art to performance art. For countless artists and movements across time, whether in support or dissent, classicism has been a fundamental point of reference.

Such opposition was still a long way off in the fifteenth and sixteenth centuries, when the trend was fresh and had yet to accumulate centuries of questionable reuses and controversial associations. In the meantime, the Renaissance penchant for ruins had one highly beneficial effect: it sowed the seeds of the modern preservation movement. In the 1500s, some critics began to lament the destruction of Rome's ruins. Early that century, Raphael, with the assistance of the humanist Baldassare Castiglione, penned an official memo to Pope Leo X, urging him to take steps to protect Rome's ancient buildings. It was a source of "great pain," he wrote, "to behold almost reduced to a cadaver this holy, noble city—once queen of the world, now so miserably wrecked." He went on to mourn the "many beautiful things [that] have been destroyed . . . many columns [that] have been broken and cracked in the middle, many architraves and beautiful friezes smashed." His words were meant as a call to action.

Why so much destruction? The answer to that question raises another Renaissance paradox: admiration for antiquity went together with its annihilation. Raphael rightly recognized that blame was not to be placed on outside invaders who had sacked Rome in the past, but rather on powerful insiders: "How many popes," he asked, "have permitted the ruin and disintegration of the ancient temples, the statues, the arches and other buildings, glory of their forebears?" The reason was simple: marble looted from ancient spots was a convenient local supply of building materials—whether simply reincorporated in a new context or melted down into lime and used for mortar. Shockingly to us today, large kilns were set up right in the Forum for that very purpose, so that the burning could happen on site, night and day, to fuel Rome's construction boom. The Renaissance far outpaced the Middle Ages when it came to this destructive practice. New St. Peter's, that glittering figurehead of urban and Church revival, rose from the ashes of hundreds of ancient buildings and sculptures.

Raphael himself is an emblem of this contradiction. On the one hand, he recognized how dire the situation was and advocated for it to change: "I would venture so far as to say that this new Rome that one sees today," he wrote to Leo X, "as grand as she is, as beautiful, as graced by palaces, churches, and other buildings, has been entirely constructed from mortar made from ancient marble." On the other hand, even as he—with Castiglione's assistance—was crafting that eloquent appeal to the pope, Raphael was acting as official architect of St. Peter's, as well as papal "inspector of antiquities": a position that allowed him to decide which buildings and objects would be re-appropriated into the splendid new church or sent to the kilns to facilitate its construction. If a move toward preservation can be detected in Raphael's time, it was the spark of an idea that would not bear fruit for centuries.

In any case, most Renaissance lovers of antiquity had grander plans. What kept them awake at night was more radical than preservation: it was the dream of a full reconstruction of the ancient city and its buildings, a complete knowledge that left no question unanswered. While Rome's golden age could not be reconstructed in reality, it could be on paper. For this reason, the sixteenth century saw the beginning of new literary and visual genres, including a new kind of map, which aimed to recreate the long-vanished city from its foundations up, including monuments real and invented. This is where the mapmakers discussed in this chapter enter the picture.

Fig. 70

Raphael, Loggia of the Villa Madama, Rome. Photo: Scala / Art Resource, NY.

We have already seen hints of this tendency in works like Leonardo Bufalini's plan, discussed in the previous chapter, which mixed the physical reality of Renaissance Rome with an idealized version of the city's past. The maps in this chapter take that impulse to a new level, purging the city's present entirely. These works are sometimes referred to as "archaeological plans," but that term does not quite fit, since it implies a modern, objective approach, geared toward a sober, accurate reconstruction. While some Renaissance mapmakers, like Marliani, were indeed motivated by this goal (even if they did not yet have the tools to realize it), others, like Ligorio, reveled in creative license. Reconstruction in the Renaissance assumed many forms. Ranging from the scholarly to the fantastical, these retrospective works show just how large the city's past loomed in the popular imagination.

The urge to restore Rome's fragmentary past through visual imagery is hardly something that ended with the Renaissance. In this chapter we will see it reach a height with Piranesi's hallucinatory vision, but it is worth noting that the urge survives today as strongly as ever. In movies and television series like *Gladiator* or HBO's *Rome*, half the pleasure in watching comes not from the narrative but from the vivid scenery, which transports us back to the brilliance and squalor of the city some two thousand years ago. Even in our own internet age, a popular tourist's souvenir is the decidedly analog genre of

"Rome-then-and-now" books, which allow readers to switch back and forth between the past and present appearance of a given site with the flip of a page. Even this time-tested crowd-pleaser, however, has recently been brought into the twenty-first century in the form of the app Atavistic.

To this list could be added many reconstructive platforms already discussed in this book—such as Rodolfo Lanciani's plan, Stanford/University of Oregon/Dartmouth's mappingrome.com, Italo Gismondi's model, and Bernard Frischer's Rome Reborn, all discussed in chapter 2. The video game *Assassin's Creed: Brotherhood* (2010) recreates Rome's historical cityscape, too, only this time that cityscape is sixteenth-century (fig. 71; compare to fig. 63). In other words, the reconstruction takes aim at the age that invented reconstruction. *Assassin's Creed* is the only game in the bunch, but in a sense, all reconstructions—with their promise of virtual time travel—share the element of play. That said, their aims are also always serious, whether they are about forcibly pulling the past into the present, writing and rewriting a distant era, or retreating into a rosy-tinged version of history.

Archaeology in Its Infancy

Bartolomeo Marliani was a scholar's scholar—known for leading his learned colleagues around Rome to share with them his considerable knowledge about the city's ancient sites. He published this map (fig. 72) as a double-page woodcut illustration for his treatise on ancient Rome, intended as a kind of erudite guidebook for his fellow antiquarians. The map is the largest illustration in the book, but it is still relatively modest in appearance—not particularly large, detailed, or adorned with the flourishes that graced other printed maps, particularly those that were separately published.

It also might take viewers a moment to get their bearings when examining it. The most evident forms are Rome's hills, which appear like lumpy, dark stones grouped about the paper, concentrated toward the top and bottom and parting lengthwise in the middle. The Tiber creeps diagonally across the map, from upper left to lower right, north to south, weaving its way through the valley that divides the hills—like a long tongue extending through a parted canine jaw.

Looking more closely, it is possible to make out some manmade features scattered across Rome's topography: the Aurelian Wall forming its irregular circuit, and within it the ground plans of several structures. In the middle of the city, near the first bend in the Tiber, a few toppled Egyptian obelisks are portrayed lying on their sides (fig. 73). To their right are the columns of Marcus Aurelius and Trajan, still upright. Little else is portrayed in the city center. If one were to think away the hills, it becomes a strangely empty cityscape.

Despite the map's rather simple appearance, it is in many ways extraordinary—for what it does and does not show, for how it shows it, and for

Fig. 72

Bartolomeo Marliani, map of imperial Rome, woodcut, from *Urbis Romae topographia*, Rome, 1544. The Getty Research Institute. Digital image courtesy of the Getty's Open Content program.

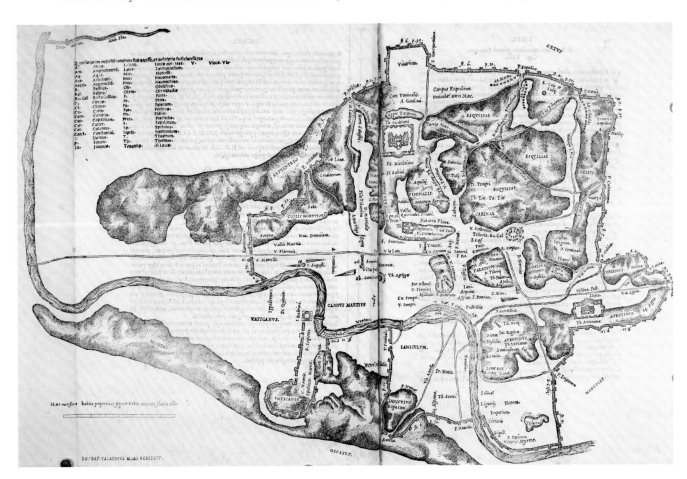

its larger implications. Marliani's map was a milestone: the earliest measured plan of Rome ever to appear in print, predating Bufalini's more elaborate version by several years. This is just one of the intriguing similarities between the two maps. Not only are both plans, but both share the same orientation, both include a scale, and both record some measurements along stretches of the Aurelian Wall. Both maps also depict the hills in a similar fashion, using cross-hatching (intersecting sets of parallel lines) in an effort to differentiate their heights from the uninked flats below. This technique can be seen as a precursor of contour lines, a convention familiar from modern topographical maps.

These parallels cannot be coincidental, and it has been speculated that Bufalini—who, unlike Marliani, had surveying expertise—helped Marliani to prepare his map, or possibly that Marliani had access to Bufalini's when it was still a work in progress. Yet there are many important differences between the maps, too. While Bufalini mixed historical eras and invented entire structures, Marliani resisted anachronism and speculation. His map portrays late imperial Rome specifically, avoiding Bufalini's timeless blend. Accordingly, the latest structure depicted is the Baths of Constantine from the early 300s. There is just one element out of time: the ninth-century Leonine Wall encircling the Vatican.

In another difference, Marliani depicted only the ground plans of buildings that had been standing in the fourth century and that still stood in his own time. Presumably, he was not comfortable representing the forms of structures unless he could inspect them with his own eyes. For this reason, just a handful

Fig. 73

Marliani, map of imperial Rome, detail showing the center, with imperial columns circled

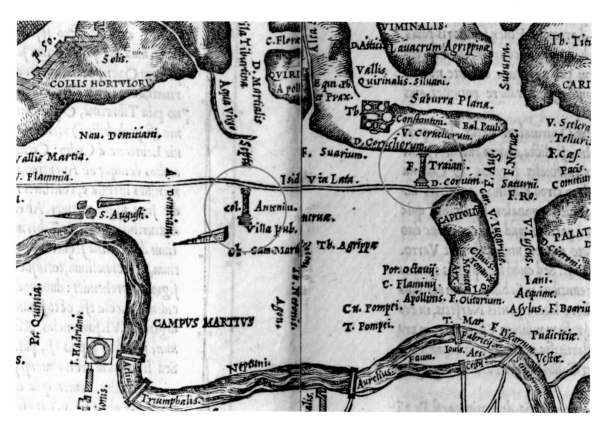

Chapter Five

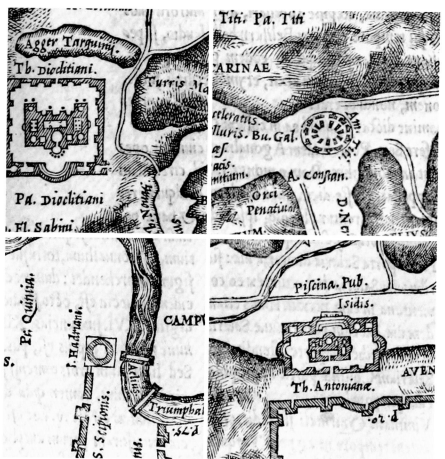

Fig. 74

Marliani, map of imperial
Rome, details showing
(clockwise from upper left)
the Baths of Diocletian,
Colosseum, Baths of
Caracalla, and Mausoleum
of Hadrian

of well-preserved monuments appear, including the Mausoleum of Hadrian, the Pantheon, and the Colosseum, as well as the bath complexes of Caracalla, Constantine, and Diocletian (fig. 74). Vanished buildings whose locations Marliani could be fairly certain of were indicated with labels only, not pictures.

The end result is a very pared down vision of ancient Rome: almost a ghost town. The map is as accurate as possible given the knowledge of the time, but it is hardly a vivid picture of the city's heyday. Instead, it effectively conveys just how much had been lost by the time the Renaissance rolled around—not only in terms of architecture but in terms of awareness about it. The ancient city comes off as distant, unknowable. For other mapmakers, that conclusion was simply unacceptable.

An Ancient Roman Theme Park

The brilliant, eccentric antiquarian scholar, painter, and architect Pirro Ligorio was born in Naples and worked his way into the most elite circles of Roman patronage. Among his projects were the jewel-box of a suburban retreat known as the Casino of Pope Pius IV in the Vatican and the superbly landscaped Villa

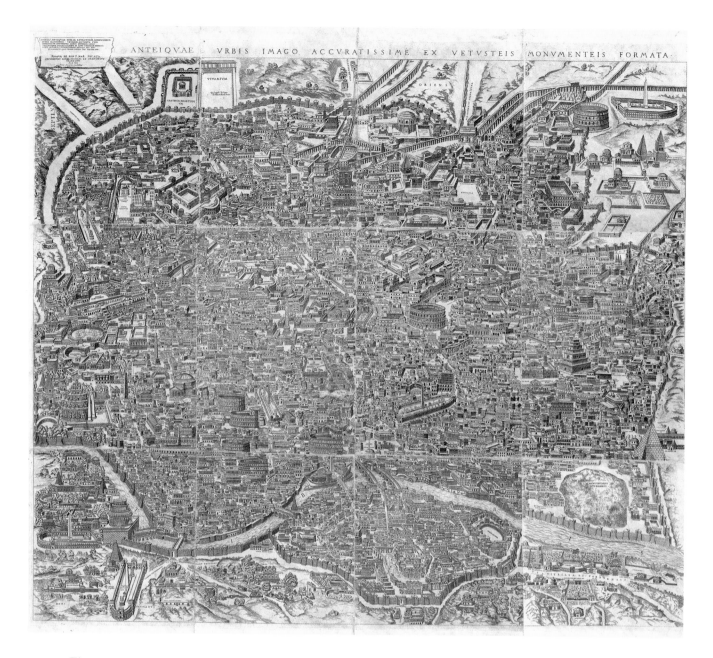

Fig. 75

Pirro Ligorio, *Anteiquae urbis imago accuratissime ex veteribus monumenteis formata*, engraving and etching, 1561 (reprint by Giovanni Scudellari, 1820–30). The Getty Research Institute. Digital image courtesy of the Getty's Open Content program.

d'Este at Tivoli, twelve miles outside of Rome, for the Ferrarese noble and cardinal Ippolito d'Este. Ligorio was also a cartographer, publishing several maps of European countries and regions, as well as a series devoted to Rome, over the course of his career.

In an age when the ancient city was a common intellectual obsession, Ligorio was unmatched in his desire to gain a profound understanding of Rome's past. This driving passion was recorded in dozens of volumes containing notes and drawings, the raw material for an ambitious publication that never materialized. In a slim book on antiquity that Ligorio did manage to publish, in 1553, he wrote that he wished "with all my heart to refresh and preserve the memory of ancient things and at the same time to satisfy those who delight in them."

Chapter Five

Nowhere did he accomplish this goal better than in the enormous and elaborate bird's-eye view of ancient Rome that appeared in 1561 (fig. 75). Before delving into the radical differences between this work and Marliani's, it is important to note that they did have something in common: both seem to depend loosely on Bufalini's survey for some of their basic topographical features, like the Aurelian Wall and the Tiber. But in every other way—visually and conceptually—they are complete opposites. While Marliani's Rome had a lot of "breathing room" on the paper, Ligorio's image is cropped close to the Aurelian Wall, and the city seems to strain at its confines. His map is many times larger than Marliani's and includes a bewildering amount of minute detail, taking the reconstructive urge to an extreme.

If Marliani was selective, Ligorio was emphatically inclusive. He portrays Rome as a crowded metropolis populated by a cornucopia of monuments, each one larger than life. There are no ruins as such to be found in his map. Instead, there are scores of intact, showy bath complexes, temples, palaces, circuses, theaters and amphitheaters, multistory apartment buildings, aqueducts, mausolea and tombs, triumphal arches, commemorative columns, and so on: a boisterous catalog of ancient buildings and building types (fig. 76). Some structures are labeled and thereby equated with actual buildings, still existing or not, but many are left anonymous, their main purpose to fill in the blank spaces with magnificence.

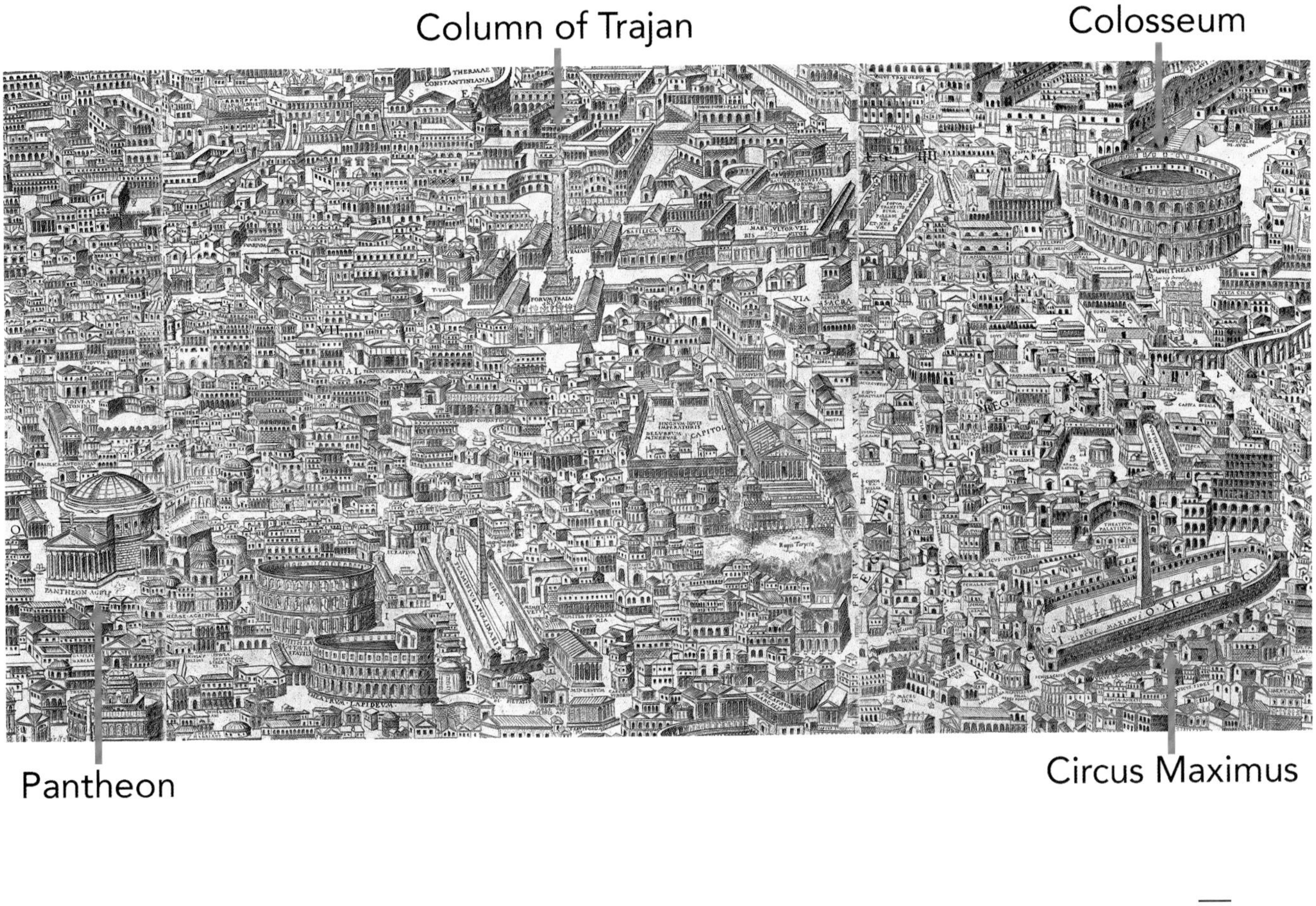

Fig. 76

Ligorio, *Anteiquae urbis imago*, detail showing the center

Column of Trajan

Colosseum

Pantheon

Circus Maximus

Fig. 77

Bronze sestertius of Trajan, 103–11 CE (reverse showing the Column of Trajan). Metropolitan Museum of Art, Rogers Fund, 1908.

For Ligorio, architectural reconstruction is conflated with architectural design. This impulse toward creativity contrasts sharply with the cautious approach we saw in Marliani's plan. Where Marliani resisted speculation, Ligorio seems to have reveled in it. This difference might be rooted in the fact that Ligorio was a practicing architect as well as a scholar. His antiquarian studies were often funneled into his own building designs, and that is the case here.

Ligorio included some recognizable landmarks, but spectators must hunt around for them in the fray. The Colosseum is probably the most obvious, and beneath it, to the left, the Pantheon can be spotted. These and other surviving buildings lend a familiar touch to what is otherwise an alien image. The vast majority of Ligorio's structures are nonexistent specters meant to evoke Rome's illustrious past. In the center, for example, a flotilla of theaters and circuses denotes this zone as one of entertainment (see below and to the right of the Pantheon). Few of these structures had left sufficient traces to really reconstruct their appearance—yet that is just what Ligorio did.

Ligorio's architectural figments might be made up, but they were not just plucked from his imagination. While many of them deviate from what archaeologists now believe to have been correct, they almost always contain a grain of truth. Ligorio was not a pure fantasist: he was quite erudite and knew his antiquities as well as any comparably well-read scholar. It was his reconstructive process that set him apart. Whereas for Marliani the absence of physical remains presented an insurmountable obstacle, for Ligorio it was just a detour. He eagerly turned to other kinds of evidence, such as ancient Roman coins (which he avidly collected), for they were often graced by little pictures of monuments (fig. 77; compare with the Column of Trajan in fig. 76). He also used the principle of analogy, drawing on the example of a known ruin to fill in a plausible form for a lost one of the same basic type.

Ligorio's map is equally fascinating for its multitude of anonymous buildings, which for the most part look like grand, exciting structures, not generic filler. His concept of the great ancient city depends on these infill structures as much as on known monuments. Through sheer abundance, he also gives a sense of the city as a living, breathing place—staying true to his goal to "refresh" its memory. The overall effect is vibrant and dazzling. Ligorio's map is not a reconstruction per se, but rather a full-fledged reinvention of ancient Rome—not as it was, but as he wished it to have been.

That said, it betrays the subtlest indications that he was either unable or unwilling to rid himself completely of his Renaissance mindset. At lower left, for example, the Circus of Nero, where St. Peter was crucified, is pictured on the Vatican Hill (fig. 78). The later presence on that site of the great basilica, however, is foreshadowed by a tiny inscription that reads, "here today is St. Peter's."

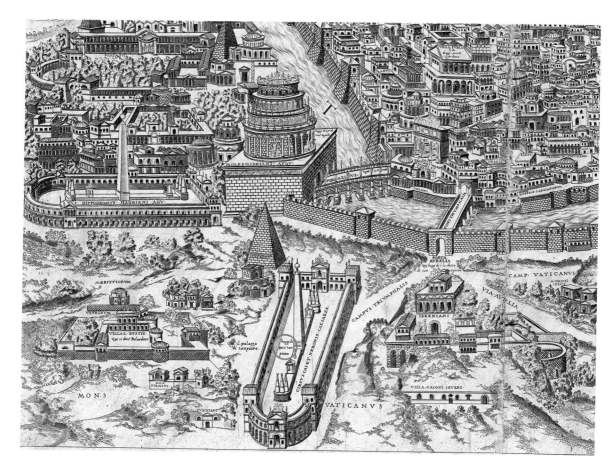

Similarly, nearby, Ligorio wrote, "here is the Palazzo of St. Peter's"—that is, the Vatican Palace. There are many of these captions scattered across the image, small but meaningful testaments to hindsight. Their insistent present tense and references to "today" pull the viewer back to the sixteenth century—a tacit admission, perhaps, that as much as Ligorio longed to immerse himself in the past, it was irretrievably lost. Or perhaps it is a more affirmative statement that the past was just a prelude to the marvelous present.

Fig. 78

Ligorio, *Anteiquae urbis imago*, detail showing the Vatican Hill, with inscription circled

A Ghostly Fantasy

Following in the footsteps of Ligorio two centuries later, the brilliant artist and architect Giovanni Battista Piranesi also conjured a breathtaking vision of a long-lost Rome. Today, Piranesi is known for his many printed views (or *vedute*) of the city, depicting its modern as well as its ancient sites. His skill and artistry were unparalleled: with his dazzling draftsmanship and etching technique, dramatic compositions, and bold contrasts of light and shadow, Piranesi has had few equals in the cityscape genre.

Piranesi was also, like Ligorio, a gifted and insightful scholar who made his own rules, as well as a cartographer. Over the course of his career, he surveyed

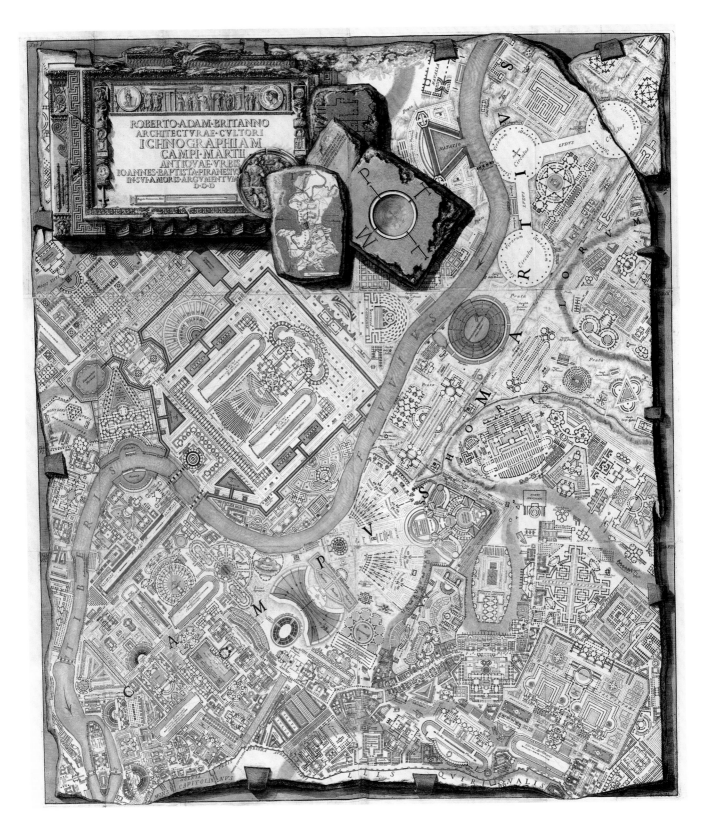

Fig. 79

Giovanni Battista Piranesi, plan of the Campus Martius,
etching, Rome, 1762. Houghton Library, Harvard University.

and issued the most accurate plan to date of Hadrian's Villa, as well as several maps of Rome that range from fairly straightforward to completely outlandish. His great map of the Campus Martius—the low-lying part of Rome nestled within the curves of the Tiber on the river's left bank—falls decisively into the latter camp.

We first met Piranesi in chapter 2, where his *Pianta di Roma* presented a restrained vision of the ancient city largely devoid of monuments (fig. 24). The only ones included were those remnants still to be found in his own time, and they were depicted in their true fragmentary state. Piranesi's criteria and sensibility in that case seem descended from Marliani's conservative method. Such was Piranesi's versatility, however, that he was able to switch gears entirely and channel Ligorio's archaeological creativity for his magnum opus, the Campus Martius. In this work, which he dedicated to his friend, Scottish architect Robert Adam, Piranesi—like Ligorio—could not bring himself to leave any ruin in ruins. Every feature is made perfect and complete. Unlike Ligorio, however, Piranesi used the measured architectural language of the ground plan to lend his ancient city an air of objective science. But science, here, is in the eye of the observer.

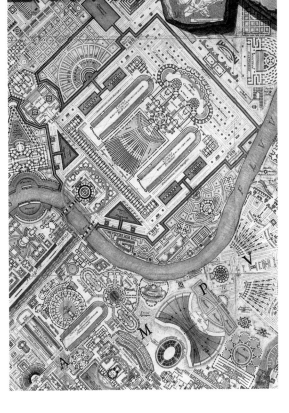

Fig. 80

Piranesi, Campus Martius, detail showing the central portion of the image, with the Mausoleum of Hadrian (above) and the Pantheon (below) circled

Like Ligorio, Piranesi included some existing monuments to help his viewers orient themselves in the otherwise thoroughly disorienting image (fig. 80). Again, survivors like the Pantheon and Mausoleum of Hadrian serve as recognizable, almost clichéd points of reference, as does the Tiber, twisting across the image from north—at upper right—to south, at lower left (where the Tiber Island can be made out; see bottom left corner of fig. 79). Toward top-center, Piranesi depicts a compass as well as a small, inset map of larger Rome to help fill in the overarching context (fig. 81). Despite these ostensibly familiarizing devices, however, the image is disconcertingly unfamiliar.

Consider the case of the Mausoleum of Hadrian, known later as the Castel Sant'Angelo: in Piranesi's map, that massive tomb is just the tiny centerpiece of a colossal, enigmatic complex—a kind of city within the city (see top of fig. 80). Here, Piranesi is thinking more in terms of contemporary urban design principles than archaeological facts. Similarly complicated, gigantic, invented compounds sprawl across Piranesi's cityscape, their facets interlocking to create an enormous puzzle that leaves little room for streets or other urban features. Their ground plans are highly geometric, symmetrical, and bizarre. In Piranesi's hands, Rome is transformed into an intricate machine.

The reconstructive purpose of Piranesi's Campus Martius is mere pretense: the map is replete with figments of his brilliant imagination. In a virtuoso illusionistic flourish, he represents the map as if inscribed upon an irregular

Fig. 81

Piranesi, Campus Martius,
detail showing the inset map
and compass (T=north)

slab of marble, attached to a wall with iron clamps. This witty touch is surely a reference to the third-century *Forma urbis*, discussed in chapter 2 (figs. 21–23). In Piranesi's day, fragments of that ancient map were displayed in just this way in the Capitoline Museum, and they shared exactly the same kind of measured architectural language. Piranesi had previously depicted bits of the marble plan at the edges of his own *Pianta di Roma* (fig. 24), the implication there that it was an irretrievably lost picture of the irretrievably lost ancient city. In his visionary Campus Martius, by contrast, he not only reconstituted but even brashly outdid the *Forma urbis*, restoring it—together with Rome itself—to an ideal, and surreal, state of wholeness.

FURTHER READING

Ames-Lewis, Francis. *The Intellectual Life of the Early Renaissance Artist*. New Haven: Yale University Press, 2000.

Barkan, Leonard. *Unearthing the Past: Archaeology and Aesthetics in the Making of Renaissance Culture*. New Haven: Yale University Press, 1999.

Chapter Five

Burns, Howard. "Pirro Ligorio's Reconstruction of Ancient Rome: The *Anteiquae Urbis Imago* of 1561." In *Pirro Ligorio: Artist and Antiquarian*, edited by Robert W. Gaston, 19–92. Florence: Silvana Editoriale, 1988.

Connors, Joseph. *Piranesi and the Campus Martius: The Missing Corso: Typography and Archaeology in Eighteenth-Century Rome*. Milan: Jaca Book, 2011.

Grafton, Anthony, Glenn W. Most, and Salvatore Settis, eds. *The Classical Tradition*. Cambridge: Harvard University Press, 2010.

Hui, Andrew. *The Poetics of Ruins in Renaissance Literature*. New York: Fordham University Press, 2016.

Jacks, Philip. *The Antiquarian and the Myth of Antiquity: The Origins of Rome in Renaissance Thought*. Cambridge: Cambridge University Press, 1993.

Karmon, David. *The Ruin of the Eternal City: Antiquity and Preservation in Renaissance Rome*. Oxford: Oxford University Press, 2011.

Pinto, John A. *Speaking Ruins: Piranesi, Architects, and Antiquity in Eighteenth-Century Rome*. Ann Arbor: University of Michigan Press, 2012.

Thompson, David. *The Idea of Rome, from Antiquity to the Renaissance*. Albuquerque: University of New Mexico Press, 1971.

Weiss, Roberto. *The Renaissance Discovery of Classical Antiquity*. 2nd ed. Oxford: Basil Blackwell, 1988.

Chapter Six

Rome of the Saints and Pilgrims

"The city, which [Martin Luther] had greeted as holy, was a sink of iniquity;
its very priests were openly infidel, and scoffed at the services they performed;
the papal courtiers were men of the most shameless lives; he was accustomed
to repeat the Italian proverb, 'If there is a hell, Rome is built over it.'"

Thomas M. Lindsay, *Luther and the German Reformation*, 1900

Even as many scholars immersed themselves in Rome's pagan antiquity, for others, it was the city's saintly side that sparked the most ardent devotion in the sixteenth century. Rome's ancient glories notwithstanding, the city's most enduring appeal was as a pilgrimage destination. Even in Rome's darkest days during the popeless fourteenth century, pilgrims continued to come. Even in the midst of the dangerously compelling critiques voiced by Martin Luther, John Calvin, and other Reformers who questioned Rome's status as a beacon to the faithful, dividing European Christians, pilgrims continued to come.

The Reformation, in fact, had a paradoxical effect, for it spurred the Roman Church out of its complacency and forced it to make itself relevant again—a renewal that ultimately caused Rome to be transformed into a showpiece, rife with glorious churches and dazzling pageantry. A whole new category of printed imagery arose during this time to publicize the city's Christian topography, catering to the lucrative pilgrimage market while countering attacks on Rome's sanctity.

The maps examined in this chapter accordingly offer up compelling pictures of Rome as a sacred place, belying the fact that its very identity as such was under fire. An image like *The Seven Churches of Rome*, published by Antonio Lafreri, equates Rome with its holy sites alone, in the process inventing a new, cartographic form of pilgrimage souvenir—one that does not even bother to include streets. Giovanni Maggi's slightly later map of the city similarly provides a catalog of Rome's major Christian shrines but groups them around a street plan to give spatial context. The longevity of these novel kinds of imagery—geared toward a specific, pious demographic and proffering a one-sided version of the city—is confirmed by their progeny, such as Tanfani & Bertarelli's map of Rome's major basilicas from some 350 years later.

Why was Rome such a draw in the first place? Yes, it had more than its share of venerable holy sites, places linked to the lives and deaths of early saints and martyrs. But Jerusalem and the Holy Land had at least as much credibility in that regard. Of course, in the Middle Ages and Renaissance, Rome was a more feasible voyage than Jerusalem—but perhaps not quite as feasible, for most Western Christians, as the pilgrimage *camino* that saw pious travelers plodding along well-established routes to traverse northern Europe and converge to the southeast, toward Santiago de Compostela in Spain. Rome, however, offered the most concentrated and accessible number of holy sites housing important relics in a single place. Most important, those spots were linked to indulgences, and by the late Middle Ages indulgences had become the stock-in-trade of pilgrims.

What were these sought-after prizes? Indulgences, simply put, were get-out-of-sin coupons, a kind of pious currency. The basic premise went something like this. All human beings were sinners, some more, some less. Certain sins—mortal ones—were so grave that they were one-way tickets to hell, unless one repented. But most sins were merely venial, meaning all one had to do to make up for them and eventually gain entrance into heaven was to confess them and be forgiven by a priest, then, later—after death—spend some time in purgatory: say, a few thousand years. And this is where indulgences came in, for they could shorten that sentence, literally for good behavior.

Indulgences had to be earned, usually through selfless acts like charity, or through codified religious rituals performed in specific locations. In Rome, more expediently than in any other city, pilgrims could gather indulgences

Fig. 82

Lucas Cranach the Elder, Martin Luther as an Augustinian monk, engraving, 1520. Metropolitan Museum of Art, Gift of Felix M. Warburg, 1920.

like squirrels gathering nuts for the winter. They did so by entering into the major basilicas at prescribed moments and/or in prescribed fashions, prostrating themselves before certain altars containing saints' relics (those above the tombs of St. Peter and St. Paul were particularly revered), and undergoing feats of physical endurance such as climbing stairs toward holy sanctuaries while on their knees in prayer.

In theory, pilgrimage was about sacrifice in the name of faith: sacrifice of one's earthly comfort and resources, financial and otherwise. Indulgences, with their promise of heavenly redemption, were the rewards for these self-inflicted hardships in the here and now. But the concept of pilgrimage had not always been so . . . transactional. Earlier in the Middle Ages, these spiritual journeys had not been geared toward reaping quantifiable returns. Their motives were purer and more personal. They had to do with achieving a perfect state of piety, feeling oneself in contact with or even at one with saints, martyrs, and Christ in the suffering he had experienced on behalf of humankind.

Indulgences—already somewhat suspect as a concept—took a darker turn in the late Middle Ages and Renaissance, when, increasingly, one could simply purchase them without taking part in the accompanying act of devotion or good deed. The traveling indulgence peddler became a familiar figure, usually, in effect, a charlatan who would come to town offering printed, stamped certificates of indulgence for purchase. Indulgence sales surged in the early 1500s, for one massive reason: they helped to finance the building of New St. Peter's Basilica in Rome.

This fact became a contentious issue for the Reformers, who saw the practice, the building that resulted from it, and the city itself as symbols of corruption

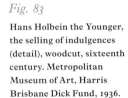

Fig. 83

Hans Holbein the Younger, the selling of indulgences (detail), woodcut, sixteenth century. Metropolitan Museum of Art, Harris Brisbane Dick Fund, 1936.

Chapter Six

and earthly materialism—in other words, of all that was wrong with the Roman Church. Luther was a particularly charismatic firebrand, and his preaching resonated with many Christians (fig. 82). "Why," thundered Luther, "does not the pope, whose wealth is today greater than the wealth of the richest Crassus, build this one basilica of St. Peter with his own money rather than with the money of poor believers?" Redemption, he felt, was not even the pope's to grant in the first place, but rather God's alone. The Swiss-based German artist Hans Holbein the Younger, an adherent of Luther, created woodcuts depicting the pope as a corrupt, morally bankrupt figurehead, in one case showing him handing out indulgences while monetary transactions take place right at his feet (fig. 83).

This matter is of more than academic importance. The promise of indulgences was one of the principal reasons why Rome continued to be a magnet for pilgrims. Jubilees or Holy Years, in particular, were designated times for Christians to come to the city precisely in order to receive such penitential pardons. For Reformers, indulgences were one among many problems with the way the Catholic Church was doing business. Other criticisms involved finer doctrinal points—but the implications were just as damning for the Church, its hierarchy, and the city it called home.

Crucially, Luther and others questioned the role of the clergy and of any mediators, such as saints, in achieving salvation. Why should priests have the power to forgive sins? Why should the faithful direct their prayers toward the saints—or, by extension, toward saints' relics—when they should be praying to Christ and God alone? The importance of performing good works was also contested. According to Luther, salvation did not come from pious deeds but instead from the grace of God alone. Earthly acts, such as charity, or pilgrimage for that matter, did not help the faithful to gain entrance to heaven (although they did not hurt, either). All of this came to be seen, rightly, as an attack on the papacy.

Luther's propositions rankled the Church on many counts. Pilgrimage was big money for Rome: the city's main industry. To discount its purifying power was to endanger the city's very livelihood and role on the international stage. Rome was rich with saints' remains, so to minimize their importance was an affront to its sacred topography. The city was Church headquarters, overflowing with clergy of many stripes, all of whom had a vested interest in upholding their own role in the redemptive process. Still, rather like the pagan emperors who had been slow to recognize the coming tidal wave that was Christianity, the Church and its vicars were slow to recognize the magnitude of the threat from the north—as though if they ignored it, it would just go away.

Of course, the Reformation did not go away. The response of the papacy and Church officials was hesitant at first, then increasingly militant. Not until almost thirty years after Luther launched his revolution did they begin to take action. In 1545, Pope Paul III—ironically, one of the most corrupt and mercenary in history—convened the Council of Trent to draft the official Church position on the Protestant criticisms (fig. 84). That assembly of theologians, prelates, experts in canon law, and other dignitaries met on and off until 1563. Their mission was to institute Church reform from within while also countering

Rome of the Saints and Pilgrims

the supposed heresies of the Reformers. Helping their cause, the religious orders proliferated, bristling with missionary zeal—not only the Jesuits but also the Theatines, Ursulines, Capuchins, and Oratorians.

In Rome, it was a conservative, even reactionary, moment for culture and the arts. Church thinkers railed against any expression or act that seemed to violate orthodox teachings. The Inquisition, charged with enforcing grave penalties against any Christian thought to have committed public heresy, swung into high gear, seeking to make examples of high-profile offenders and riffraff alike. Galileo Galilei famously ran afoul of this group of hardliners, buckling under the threat of physical torture for daring to question conventional doctrine regarding the cosmos. As the end of the sixteenth century neared, however, the tide started to turn—not the Inquisition, which kept up its repressive practices for years (indeed, Galileo's trial occurred in 1633)—but rather the climate in Rome. Perhaps because the state of the city and of the Church seemed to have stabilized, the tone lightened, and the general sense of being under siege lifted.

At the same time, Church patrons realized that they could use cultural expression as a weapon in their arsenal: to strengthen the faith, stem the flow of Christians, and even win new converts. Awakening from the prevailing somber and cautious mood, artists and architects embraced drama, dynamism, even sensory overload. This ethos is the essence of the Baroque period, known for its theatrical art of persuasion. Caravaggio, Bernini, Borromini, and others exemplify this turn toward emotional exuberance. Baroque Rome was a heady place,

Chapter Six

and an exciting one. Its cultural momentum seemed to fly in the face of Protestants who had likened the city to Babylon and predicted—or longed for—its demise. Rome was reborn, yet again, as center and flagship of the Catholic world.

Pilgrimage soared during this time. The Holy Year of 1575 attracted some 400,000 pilgrims—a tenfold increase over previous Jubilees—while that of 1600 attracted more than 500,000: over five times the permanent population of Rome at the time. This period also saw a corresponding rise in the number of images, including maps, created as souvenirs for that demographic. New printed genres arose to proclaim the sanctity of the Eternal City and its centrality in Christian devotion. If the maps examined in the previous chapter were created by and for scholars interested solely in Rome's antiquity, their limited, backward-looking focus almost seems a willful turning away from the current atmosphere of religious contention. The maps we turn to now, by contrast, were geared toward pious visitors, for whom the city's pagan monuments were just a sidetrack—a curiosity. Their sights were trained on Rome's glorious bounty of holy places and on the earthly and heavenly benefits they offered.

The Way of the Faithful

It would be hard to overstate the influence of this map (fig. 85), published by Roman print impresario Antonio Lafreri and designed, it is believed, by Stefano Du Pérac or possibly Giovanni Ambrogio Brambilla. The image spawned scores of imitations in the decades and centuries after it first appeared, becoming the prototype for a new category: the "seven churches" map. What made Lafreri's prototype so successful? In short, it was a perfect encapsulation of the goals, ideals, and experience of Roman pilgrimage.

Lafreri's map was intended to coincide with the Jubilee of 1575, and thereby tap into the influx of pious visitors eager to purchase an image commemorating their time in Rome. Although they lasted a full year, Jubilees bore certain parallels to the Olympics in modern times: they occurred at defined intervals, stimulating new construction as well as renovation, infrastructural development as well as maintenance, all in hopes of drawing legions of visitors from near and far—and profiting from them. Projects like the Ponte Sisto and the network of streets of Sixtus V, we have seen, responded to a dual imperative: benefitting pilgrims in the shorter term and permanent residents in the longer term. The city's great basilicas, which housed Rome's holiest relics, also owed much of their upkeep and restorations to their status as major attractions.

Jubilees were also catalysts for map production—but not because pilgrims used them to find their way around. In the Renaissance, visitors navigated the city by asking directions, following the crowds, or hiring human guides. Maps like Lafreri's were not practical tools but rather desirable mementos to bring home. Printed images of Rome that vaguely filled this niche had been available for the better part of a century before Lafreri's map appeared, but none had so strategically pinpointed the aspects of Rome that appealed to pilgrims, and as a result, none had been so popular with that target audience.

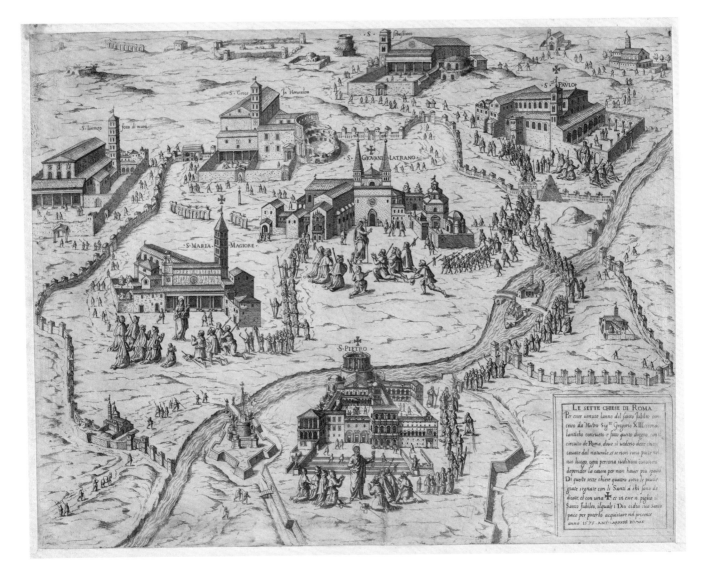

Fig. 85

Le sette chiese di Roma,
engraving and etching,
Rome: Antonio Lafreri, 1575.
Metropolitan Museum of Art,
Rogers Fund, Transferred
from the Library, 1941.

Lafreri's map offers a kind of CliffsNotes version of pilgrimage. In the text at lower right, he specifically refers to the Holy Year of 1575 and to the necessity of visiting the city's chief basilicas in order to obtain the Jubilee indulgence. The holy grail of indulgences, so to speak, was a plenary indulgence, which forgave a repentant sinner's entire purgatorial sentence. Boniface VIII inaugurated the tradition of offering such generous remittance to pilgrims coming to Rome when he instituted the first Holy Year of 1300. To earn it, pilgrims had to spend fifteen days visiting the basilicas of Peter and Paul (twice as long if they were from Rome, since they had not had to undertake the journey to get there). Subsequent popes added St. John Lateran and Santa Maria Maggiore to the list, with these four together comprising the major basilicas of Rome. By 1575, three more had become part of the conventional circuit: San Lorenzo fuori le Mura (or St. Lawrence Outside the Walls), Santa Croce in Gerusalemme, and San Sebastiano fuori le Mura (or St. Sebastian Outside the Walls).

Chapter Six

Lafreri's map turns a spotlight on these very churches, depicting them as isolated features rising from a token cityscape. Rome is glimpsed in a bird's-eye view from the northwest. The Aurelian Wall snakes around the image, demarcating the city's physical limits, but not its sacred limits, since several of the basilicas lie outside its circuit. The Tiber with its island is shown curving through the city from lower left to upper right, but none of Rome's famous hills appear, and natural topography as such is all but nonexistent. There is also no sign of the network of streets, or any urban infrastructure to speak of. The image—and by implication the city—is truly all about the churches.

St. John Lateran, Rome's cathedral and first seat of the popes, is at the center, with St. Peter's just below it on the central axis, its unfinished dome rising above the nave of the old basilica, still standing. Both churches are turned to put their best face forward, as are all the others pictured: Santa Maria Maggiore at left, San Lorenzo above it, then—moving clockwise—Santa Croce in Gerusalemme, San Sebastiano, and St. Paul's (San Paolo). Even though streets are absent, invisible paths through the city are suggested by the pilgrims marching resolutely from one basilica to the next. The designer of the image has simply removed all obstacles—any hint of the real messiness of Renaissance Rome—in order to facilitate their travel between holy spots. In this way, the image captures the mindset that would give birth, just a decade later, to Sixtus V's revolutionary new streets connecting the city's major basilicas.

Fig. 86

Le sette chiese, detail showing pilgrims praying to St. John

Fig. 87

Le sette chiese, detail
showing hazy ruins toward
the top of the image

In the image, scattered figures can be seen entering the city at Porta del Popolo, at lower left, to begin their tour. They seem to come together and gain steam as they move among the four major basilicas, growing in number and coalescing into orderly processions. This activity corresponds to historical fact: during the Jubilee of 1575, for the first time, processions led by lay religious brotherhoods known as confraternities became a common feature of Jubilee activities. Pilgrims in the past had tended to move rather haphazardly from one spot to another in small groups of friends or family, then participate in more formalized rites within the basilicas. In 1575, the very act of traversing the cityscape became more ritualized and choreographed. In return for participating in these large processions, which sometimes involved thousands of people, pilgrims were granted a reduction in the time required to complete their indulgence requirements—from fifteen to just three days.

The image is also defined by selective manipulations of scale. Not only are the churches magnified relative to the cityscape, but the human figures also grow in size, their proportions increasing as they congregate before the main sites. In front of the four chief basilicas, pilgrims kneel reverently before standing personifications of the holy namesakes. Peter, John, Paul, and Mary stand beneficently before their respective churches, towering over their devotees (fig. 86). These figures are meant to signal the presence of their own holy relics within their respective churches: the physical traces that were the true lodestone of each place, promising salvation. Although seldom seen—for relics tended to be revealed only on special feast days—they were the primary reason for Rome's enduring appeal.

Lafreri's map presents a one-sided view of the city, but it would be wrong to suppose that pilgrims had blinders on when it came to Rome's ancient monuments. Popular medieval guidebooks known as the *Mirabilia urbis Romae* (or *Marvels of Rome*) are filled with legends about the ruins, often inflected with a Christian twist having to do with the triumph of the true faith over paganism. In the Renaissance, they continued to draw attention, and not just among scholars. But their lesser importance to pilgrims is symbolized in Lafreri's map by the presence of a few ruins that appear toward the horizon that is just squeezed in at the top of the image (fig. 87). Hazy as they appear, these are specific as opposed to generic ruins, including the Tomb of Cecilia Metella and the Circus of

Maxentius. Still, they are secondary to the map's main focus. By accentuating the seven churches, the map attests to the transformative power of pilgrimage in general and of pilgrimage to Rome—with its amazing array of supernatural relics—in particular. Coming right in the midst of the Counter-Reformation, the image was an implicit, but forceful, rejoinder to critics, yet it quickly became an icon that transcended the specific historical circumstances.

Scenes from a Pilgrimage

Giovanni Maggi produced his own take on the seven churches theme for the subsequent Jubilee of 1600. It too proved immensely successful, rivaling the popularity of Lafreri's image. New, updated editions were published for the subsequent Jubilees of 1625 and 1650 (the latter being illustrated here in fig. 88), and in the meantime the map was also sold to the many pilgrims who visited Rome on an ongoing basis. Versions of this map appeared as late as 1750. Maggi's print was roughly the same size as Lafreri's, and both were considerably smaller than, say, Tempesta's view or Bufalini's plan. Their reduced dimensions mean that

Fig. 88

Giovanni Maggi, *Descriptio urbis Romae novissima*, engraving, 1650. The Getty Research Institute. Digital image courtesy of the Getty's Open Content program.

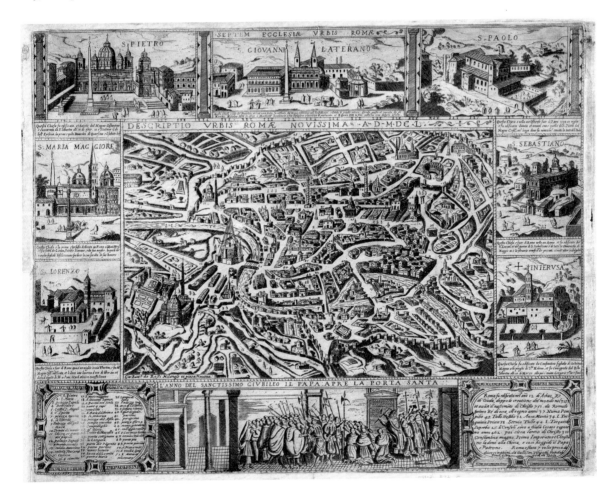

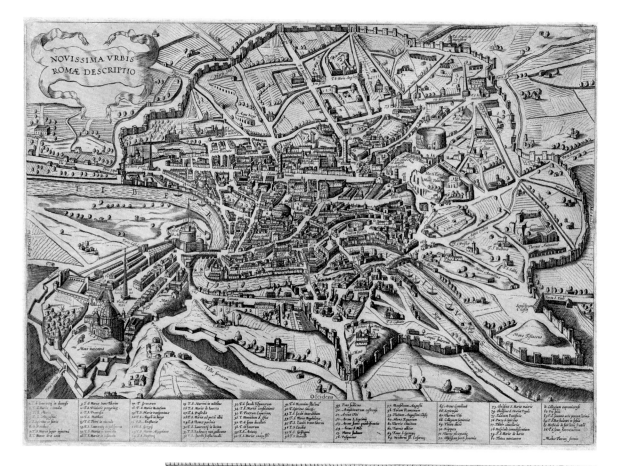

Fig. 89

Above: Matteo Florimi, *Novissima urbis Romae descriptio*, engraving and etching, 1590s. The Getty Research Institute. Digital image courtesy of the Getty's Open Content program.

Fig. 90

Below: Maggi, *Descriptio urbis Romae*, vignette depicting the opening of the Porta Santa

they were less expensive. Lafreri and Maggi were courting the pilgrimage market, aiming for a high volume of sales to a wider spectrum of collectors.

Maggi's view is more cartographic than Lafreri's, in the sense that he placed a fairly detailed street plan at the center, grouping pictures of the principal churches and a vignette relating to the Holy Year around the edges. Given the inclusion of the plan, it would seem perfectly reasonable to assume that this image, finally, was meant to be used by pilgrims to navigate Rome's warren of streets. But there is one problem with that interpretation. For the central portion of his image, Maggi was copying an earlier map by Matteo Florimi (fig. 89). In order to make room for the pictures he wished to add around the sides, Maggi was obliged to crop the edges of Florimi's map, limiting it to the city center. As a result, only three of the seven churches actually appear on the street plan.

Chapter Six

How useful could it have been? Rather than wayfinding, therefore, the map's main purpose seems to have been to provide a general sense of urban context. Like so many other printed maps until the eighteenth century, this image was a keepsake, not a practical tool.

The commemorative nature of Maggi's print is signaled by the vignette at the bottom depicting the pope opening the Porta Santa, or Holy Door, at St. Peter's (fig. 90). This entrance sits on the far right as one faces the church from the vestibule. To this day, it is open only during Jubilee years. At all other times it is sealed from within by brick and mortar, and its ceremonial opening by the pope wielding a hammer marks the inauguration of a given Holy Year. (A Renaissance commentator, Giovanni Rucellai, wrote of the tradition, describing how pilgrims who were present for the ritual breaking would snatch a fragment of the shattered wall to take away as a memento, treating it almost like a holy relic in itself.) Since pilgrims had to pass the threshold of the Porta Santa to gain the plenary indulgence associated with the Jubilee, this image held special significance to anyone who had come to Rome for that purpose.

The vignette and street plan at the center of the image notwithstanding, the star attractions of this image are Rome's holy basilicas. The three most important and earliest are depicted in insets across the top. At left is the premier pilgrim's destination of the city: St. Peter's (fig. 91). The picture documents changes to the building and its immediate surroundings since Tempesta had finally shown it complete with its dome. In the decades since that time, the remains of Old St. Peter's and its ancillary structures had been cleared away, creating a somewhat irregular open space in front, centered around the great

Fig. 91

Maggi, *Descriptio urbis Romae*, vignette of St. Peter's

Fig. 92

Maggi, *Descriptio urbis Romae*, vignette of St. Paul's

obelisk that had been placed there by Sixtus V. Gianlorenzo Bernini's grand piazza, so familiar today, was still a few years in the future.

By contrast, the new façade that had been constructed in the 1610s according to a design by Carlo Maderno is in full evidence. Maggi's picture includes a falsehood, however: namely the two bell towers shown capping each end of the façade. Maderno's plan had called for these features, and their anticipated construction was cause for them to be added to the 1625 edition of Maggi's map, despite the fact that they existed on paper only. In the late 1630s, the task of erecting them was entrusted to Bernini, golden boy of Baroque Rome's vibrant artistic culture, favorite of a whole series of powerful cardinals and popes.

Despite warnings that the foundation of the façade would not hold under the weight of these massive features, Bernini persevered, overseeing construction of the first, southern bell tower. Before he could begin work on the northern tower, the façade began to crack from the weight. In 1646, the construction that had been completed was demolished to prevent further damage. This moment was the single major public disgrace of Bernini's otherwise charmed life and career. The 1650 edition of Maggi's map negates the sad end of the episode, failing to remove the ill-fated architectural features and thereby to represent St. Peter's in its actual midcentury state. This tale is a good reminder that wishful thinking sometimes works its way into maps—and that they sometimes knowingly perpetuate falsehoods.

Mirroring St. Peter's at right is St. Paul's Outside the Walls (fig. 92). Through the Middle Ages and into the Renaissance, these two basilicas were considered a matched pair, just as their namesakes were (and are) the twin patrons of Rome. Peter was the apostle to the Jews, while Paul, a Roman citizen, was the apostle to the Gentiles. Although today the basilica dedicated to Paul has faded in prominence relative to its more famous counterpart, it was originally every bit as glorious as that built in honor of Peter. Moreover, unlike St. Peter's and other early basilicas in Rome, St. Paul's survived intact, without major renovations to its form, for some 1,500 years. In fact, it was still holding strong until the tragic fire of 1823 brought down its roof and led to its collapse. The basilica that pilgrims and tourists are able to visit today might be an impressive reconstruction, but it is also in many ways a sanitized version of the church that stood previously on the site.

On Maggi's map, St. Paul's appears as a fairly conventional, large basilica with a central nave raised above lower side aisles, crossed by a transept arm to create a Latin-cross plan. In front of the church proper is its atrium, or forecourt, a typical early Christian feature that had also been present at Old St. Peter's. Surrounding St. Paul's are structures that had grown up around it over the course of the Middle Ages, including a monastery and remnants of a fortified citadel. These accretions signal that buildings like St. Peter's or St. Paul's, in order to enjoy such extraordinarily long lives, had to become living history. Like Rome itself, that is, they needed to resist stagnation through growth and change.

Taking pride of place in the top-central panel of Maggi's map, between the basilicas dedicated to Peter and Paul, is Saint John Lateran: the cathedral of Rome and first seat of the popes (fig. 93). Like all the early basilicas, it appears as a patchwork of elements from more than a thousand years of history. The core of the church dated from the early fourth century, while its arcaded north façade as pictured and the adjoining papal palace were added by Sixtus V in the 1580s. The same pope's signature obelisk dominates the space in front, while at right is the octagonal baptistery built more than a thousand years earlier, in the fifth century. The Lateran Basilica did not share the distinction of housing apostolic tombs, but together with the chapel of the Sancta Sanctorum (or Holy of Holies) in the adjacent palace, it held some of the city's most highly venerated relics, including the heads of Peter and Paul; wood from the table of the Last Supper; and the staircase of Pontius Pilate, which Christ had ascended on his way to being sentenced to death, taken from the Holy Land.

This trio of basilicas, together with the four additional ones pictured at the sides of Maggi's map, were the jewels in the crown of Rome's sanctity. By the mid-1600s, print publishers had begun to capitalize on the market for images that interwove these sites with the fabric of the city itself. Rome, for its part, had not just stabilized since the upheavals of the sixteenth century, but even

Fig. 93

Maggi, *Descriptio urbis Romae*, vignette of St. John Lateran

flourished. The city was a source of fascination for its pagan and Christian heritage alike. As such, it was perfectly poised to become a great tourist mecca in the following centuries.

A Pilgrimage Map for the Modern Era

Dating from three centuries after the images by Lafreri and Maggi, this anonymous map (fig. 94) is in key ways their descendant. Produced for a Jubilee year, its main focus is Rome's sacred sites, its intended audience pious visitors. Rome is depicted in an aerial view from the north, the "pilgrim's perspective" familiar from the Middle Ages. The four major basilicas are singled out as pop-up pictures on the flat street plan—St. Peter's at lower right, St. Paul's above it on the horizon, St. John Lateran at upper left, and Santa Maria Maggiore below it (fig. 95). All are depicted disproportionately large and highlighted in red, which is also used to indicate the locations of the city's many parish churches, identified in a legend at right.

Yet the map also differs from its ancestors in meaningful ways. First and foremost, it was meant to be useful in a way they were not. The example reproduced here was well loved, and it bears the battle scars to prove it. Folded into a booklet for portability, the map's thin, yellowed paper is worn out at certain

Fig. 94

Tanfani & Bertarelli, *Roma: Le quattro S.S. Basiliche e le parrocchie: Anno santo straordinario MCMXXXIII–MCMXXXIV,* chromolithograph, 1933–34. National Library of Poland, via europeana.eu.

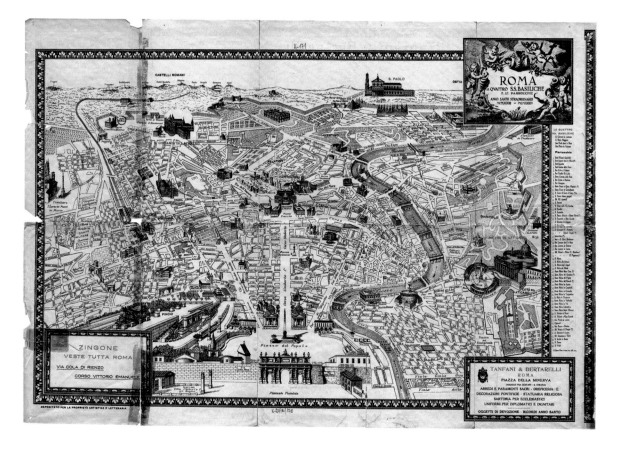

Chapter Six

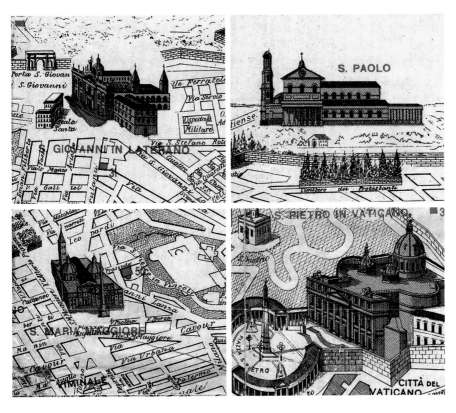

Fig. 95

Tanfani & Bertarelli,
Roma, detail showing the
four major basilicas

stress points as well as torn in places, while an adhesive has stained some of
the creases. It is, moreover, a chromolithograph (an inexpensive color printing
technique associated with mass production), characterized by relatively low
production values.

The map was made for Tanfani & Bertarelli, a Roman purveyor of religious
vessels and jewels—chalices, rosaries, etc.—whose advertisement appears at
lower right as well as on the cover of the slip case. An ad for another vendor, the
retailer of religious apparel Zingone, appears at lower left, suggesting that they
too underwrote the printing. The map was presumably produced for these establishments to distribute as a freebie, or perhaps to sell for a nominal fee. Not
coincidentally, both businesses helpfully signal their locations on the map in
red (both had also cosponsored an earlier version of this map for the previous
Jubilee of 1925). All of these signs point to it being less of a keepsake than a
utilitarian, disposable item.

The map is just as interesting as its ancestors, however, from a historical
point of view, for it reflects a fraught moment in the history of Rome and the
Church. It was produced for the special Holy Year of 1933–34, nominally declared in celebration of the 1,900th anniversary of Christ's Passion (the events
leading up to and including his death and resurrection). But there was a subtext.
The papacy had recently resolved its decades-long dispute with the Italian government regarding sovereignty over Rome—the so-called "Roman Question."
The Lateran Treaty of 1929, signed by Fascist dictator Benito Mussolini and
Pope Pius XI, had formally declared Vatican City an independent state.

This concession was a consolation prize. Almost six decades earlier, in 1871, Rome had been declared capital of recently united Italy, which resulted in the papacy losing all temporal power over the city as well as the territorial holdings collectively known as the Papal States. The 1933–34 Jubilee was the most elaborate and festive in more than a century, a major public proclamation of reconciliation between the papacy and the Fascist government of Italy. Hundreds of years after the Reformation and Counter-Reformation, this map serves as a good reminder that church and state have never really been separate in Roman affairs and that pilgrimage can carry a potent political charge.

FURTHER READING

Balbi de Caro, Silvana, ed. *Roma tra mappe e medaglie: Memorie degli Anni Santi*. Rome: Libreria dello Stato, 2015.

Birch, Debra J. *Pilgrimage to Rome in the Middle Ages: Continuity and Change*. Woodbridge, UK: Boydell Press, 1998.

Fagiolo, Marcello, and Maria Luisa Madonna, eds. *Roma sancta: La città delle basiliche*. Rome: Gangemi, 1985.

Higginson, Peter. "Time and Papal Power: The Pilgrim's Experience of the Old and New in Early Modern Rome." In *The Enduring Instant: Time and the Spectator in the Visual Arts*, ed. Antoinette Roesler-Friedenthal and Johannes Nathan, 193–208. Berlin: Gebr. Mann Verlag, 2003.

Marigliani, Clemente, ed. *Le piante di Roma delle collezioni private*. Rome: Provincia di Roma, 2007.

McPhee, Sarah. *Bernini and the Bell Towers: Architecture and Politics at the Vatican*. New Haven: Yale University Press, 2002.

Robertson, Clare. *Rome 1600: The City and the Visual Arts under Clement VIII*. New Haven: Yale University Press, 2015.

Wisch, Barbara. "The Matrix: 'Le Sette Chiese di Roma' of 1575 and the Image of Pilgrimage." *Memoirs of the American Academy in Rome* 56/57 (2011/2012): 271–303.

Chapter Six

Chapter Seven

Rome of the Grand Tourists

"Now, at last, I have arrived in this first city of the world! . . . All the dreams
of my youth have come to life; the first engravings I remember—my father
hung views of Rome in the hall—I now see in reality . . . everything is just
as I imagined it, yet everything is new."

Johann Wolfgang von Goethe, *Italian Journey*, 1786

In the 1600s, the pilgrims who flocked to Rome began to be rivaled by a new
breed of elite secular travelers: the Grand Tourists, patrician young men (and
occasionally women) who came from points north of the Alps, especially Great
Britain, to complete their formation with a strong dose of culture and history in
situ. Rome was just one stop on their Italian itinerary, but for many it was the
most anticipated, and well-to-do visitors often lingered in the city for months
and more. Grand Tourists had money to spend and a ravenous appetite for col-
lecting anything and everything relating to their travels: painted city views, or
vedute, and maps, as well as printed views (Piranesi's etchings were especially
popular), illustrated books, objets d'art, antiquities, scientific instruments, geological specimens, and so on.

Giovanni Paolo Panini was a quintessential Grand Tour artist, whose *vedute*
and creative combinations of Italian scenery (or *capricci*) were perfectly tailored

to his clients' desires. His somewhat allegorical portrayal *Ancient Rome*, together with its pendant *Modern Rome*, captures the sense of the city and its treasures as a huge marketplace to be consumed (figs. 96 and 97). The painting shows a sumptuous classical hall filled with famous antiquities and painted versions of the city's older sights lined up as if for sale, like so many bonbons for the taking. Eager, discerning, aristocratic shoppers mill about, with the patron of the work at center holding a guidebook and staring out at the viewer. The painting flatters the collector by showing him as a person of means, taste, and class, while packaging a splendid version of the city that could be acquired and controlled. The reality was, as always, quite different from the ideal, but then again consumerism is never really about reality.

In any case, not every Grand Tourist was in the market for a Panini, or for that matter a Canaletto or Batoni, and the Roman print industry expanded dramatically in the late seventeenth and eighteenth centuries to fill this important niche. Much as they did for the pilgrims, an increasing number of enterprising publishers catered to these visitors' desires for visual mementos to hang on their walls after their journey had come to an end (or, as Goethe's words, quoted above, suggest, to prepare them mentally well before it began).

This competitive atmosphere stoked an amazing array of images, from small illustrations contained in books that detailed journeys through the city, to midsize maps that began to move in the direction of practical on-site use—like François Nodot's early tourist plan discussed below—to large and glorious statement pieces, meant to hang on a wall and impress anyone lucky enough to see or own them. In a sense, magnificent items like those we will examine in this chapter by Giovanni Battista Falda, Giambattista Nolli, and Giuseppe

Fig. 97

Giovanni Paolo Panini,
Ancient Rome, oil on canvas,
1757. Metropolitan Museum
of Art, Gwynne Andrews
Fund, 1952.

Vasi were meant as stand-ins for the city itself. Like Rome, they were imposing in size, brimming with visual interest and delight and offering the promise of cultural riches. What is more, they did all this in two dimensions.

The Grand Tour stemmed in part from the earlier phenomenon of artistic pilgrimage. Northern European artists had a longstanding tradition of traveling to Italy to soak up their colleagues' innovations and to study works of the past. Albrecht Dürer spent time in Venice, while Marten van Heemskerck and other Netherlandish artists embarked on arduous but formative journeys to Rome and other Italian cities. Heemskerck and his contemporaries were followed in the 1600s by painters the likes of Pieter Paul Rubens and Nicolas Poussin, the latter of whom ended up spending most of his life and career in the city. The notion that a sojourn in Rome was vital to artistic development was formalized later in the seventeenth century when French painters, sculptors, and eventually architects competed on a yearly basis for the Prix de Rome: a prestigious prize from the state-sponsored fine arts academy financing several years of independent study in the city.

Meanwhile, the notion that such a journey was a key coming-of-age ritual caught on with the wealthy upper classes more generally. In 1796, Edward Gibbon commented, in his memoirs, "According to the law of custom, and perhaps of reason, foreign travel completes the education of an English gentleman." Most of these visitors were amateurs in the true and best possible sense of the word, which is to say lovers of art and culture. An increasing number, however, set out on a Grand Tour because it was seen as a mark of class: an obligatory rite of passage for anyone of a certain social rank. In either case, unlike their artist predecessors, they had the means to travel in style.

Rome of the Grand Tourists

What Roman sites did Grand Tourists wish to see? The answer lies in Panini's two canvases of ancient and modern Rome, specifically in the paintings-within-the-paintings. The interests of these affluent visitors overlapped to a certain extent with those of artists and of pilgrims. Many were drawn to the early Christian basilicas and martyrdom sites that attracted their devout counterparts, but the scarcity of such places in Panini's two paintings betrays the fact that Grand Tourists were not particularly driven by piety. Indeed, many were Protestant, so Roman Catholicism was, if anything, beside the point. For them, what made a site meaningful—and worth a visit—was its (secular) historical and artistic value.

As Panini's *Ancient Rome* suggests, the prime focus of Grand Tourists was classical: they set their sights on the Roman and imperial fora, the Capitoline and Palatine hills, ruins of the Colosseum, imperial bath complexes, and so on. That Panini paired his *Ancient Rome* with *Modern Rome* additionally reflects the growing importance placed on the city's Renaissance monuments, many of which had come to be considered "new" classics. *Modern Rome* also includes a fair share of very recent additions to the cityscape, such as the Spanish Steps, designed by Francesco de Sanctis and completed in 1725. Indeed, visitors marveled at the excitement and splendor of Baroque Rome, which was bustling with ambitious urban and architectural projects—truly a place to see and be seen (fig. 98).

Grand Tourists, like the artists who came before them, were also drawn by the gravitational pull of individual masterpieces. Rome was and is an open-air museum, so the city's fountains, for example, offered a feast for the eyes along with fresh drinking water. Fontana's Fountain of Moses from the 1580s is pictured in Panini's *Modern Rome*, as is Bernini's Fountain of the Four Rivers from the 1650s. Other sculptures and paintings were readily accessible in Roman churches, as they are to this day. Examples include Michelangelo's *Pietà* and *Moses* (the latter of which Panini conveniently installed in his *Modern Rome* gallery), Caravaggio's *Calling of St. Matthew*, and Bernini's *St. Teresa in Ecstasy*.

Public museums, however, were still largely in the future—or only in their infancy toward the later part of the Grand Tour era. Rome happens to be home to two of the earliest in the world, namely the Capitoline Museum and the Vatican sculpture collection, but otherwise the concept of an art collection open to all did not gain traction until the late eighteenth century. Prior to that point, visitors were able to access Rome's stellar private collections only if they had friends in high places. This is one reason why tourism in Rome and elsewhere remained the exclusive province of the privileged classes for as long as it did.

As the Grand Tour gained steam in the late 1600s, new literary genres grew up around the ritual of travel that pointed forward to the beginnings of the popular guidebook tradition in the nineteenth century. One common form was travelogues: published reminiscences of aristocrats who toured Europe. Richard Lassels's 1670 *Voyage of Italy* was one of the earliest and included a wealth of practical information, but the best-selling was Thomas Nugent's *Grand Tour* (1749), and perhaps the most famous remains Goethe's *Italian Journey* (1786). As ostensible memoirs, these were very different from the impersonal and informative guidebook genre that would emerge later. While many did offer tips for places to go and things to see, as well as suggested itineraries, they almost never included maps or illustrations to help orient their readers visually or spatially.

Over the same time span, more practical volumes began to appear in increasing numbers. In 1722, Jonathan Richardson and his son Jonathan Richardson Jr. published their *Account of Some of the Statues, Bas-Reliefs, Drawings, and Pictures in Italy*, which provided a kind of checklist for enterprising travelers. Giuseppe Vasi's *Itinerario istruttivo di Roma antica e moderna* (or *Instructive Itinerary to Ancient and Modern Rome*), first published in 1763 and then in many subsequent editions, took utility a step further. A portable pocket guidebook, it outlined eight daylong walking tours of Roman sights, describing the history and importance of the landmarks. This book served as a guide *and* a keepsake, especially if—as Vasi hoped—readers bought one of his other, grander publications to go with it: his ten-volume, lavishly illustrated *Magnificenze di Roma*, or the huge bird's-eye view known as the *Prospetto* or *Prospect* (see fig. 110).

Vasi's *Itinerario* differed from modern guidebooks in that it had few illustrations and, despite its spatial organization, did not include any maps. It is hard to imagine using it without additional assistance to navigate the city. In this era, however, we do begin to detect signs that maps were being used for on-site consultation. Sometimes, as we will see, the maps themselves suggest their practical function, but there is external evidence, too. Pompeo Batoni's depiction of Francis Basset, one of many similar "swagger portraits" the Roman artist painted for English noblemen to commemorate their Grand Tour, shows the young, fashionable subject leaning with studied, casual elegance against a pedestal adorned by a classical relief (fig. 99). In his left hand, he holds a partially rolled map of Rome, unfurled just enough to show the Vatican and the northern section of the city peeking out (fig. 100).

That very same area is visible in the flesh behind him, where St. Peter's and the Castel Sant'Angelo appear conspicuously in the distance. The background is reminiscent of a painted stage set, or an artificial backdrop pulled down like a curtain. The whole portrait is artfully staged to portray a worldly, confident young man who has mastered the art of sophisticated travel and has the great sites of Western civilization at his fingertips. The map in his hand is a prop, but a telling one, for the implication is that Basset has used it to make sense of his environment—even to gain mastery over it (which he accomplishes by going out into the world, not making it come to him, as was the case with Panini's patron in *Ancient* and *Modern Rome*). However contrived and formulaic, Batoni's portrait must hold a grain of truth with regard to the changing function of maps.

For the most part, however, maps still had not caught on as navigational tools. The modern sightseeing plan as such had not yet been invented: as we will see in the next chapter, it truly came into being along with mass tourism. In any case, Grand Tourists did not really need such practical items, since they typically employed the services of professional, learned guides, known as *ciceroni*, to accompany them around the city. Many of these figures can be seen gesticulating to their charges as they gaze up at grand ruins in the elite souvenir *vedute* paintings of the time by Canaletto and others (fig. 101).

Like those paintings, and like the seven churches prints geared toward pilgrims, the majority of maps produced for Grand Tourists were glorified mementos, meant to distill and preserve a certain experience of the city. In this sense, they also operated in a manner akin to Batoni's portraits, which—although personalized—were created for exactly the same clientele: aristocratic British travelers seeking tangible tokens of their journey to Rome.

Maps for Grand Tourists were issued in a variety of formats, but the largest were costly prestige items, intended for study, admiration, display, and reliving the Eternal City from a distance of time and space. Almost all the following images—which include the most spectacular printed maps to emerge from the Grand Tour era—fall into this general category. Of course, they also give off an unmistakable whiff of conspicuous consumption. If many Grand Tourists were motivated by social and class expectations as much as by genuine passion for the Renaissance and antiquity, so too the maps they acquired and showed off at home were meant as status symbols: testaments to the worldliness, taste, and

resources of their collectors. They took their place among a wide spectrum of items that Grand Tourists eagerly amassed during their travels.

The Grand Tour took flight in the mid-seventeenth century, after the 1648 Treaty of Westphalia ended a century of war among European states, and was curtailed in 1796, when Napoleon invaded Italy, setting off a period of war that would last two decades. Aristocratic travelers continued to come for lengthy periods after that, to be sure, but the Grand Tour's cachet was diluted following the advent of popular tourism in the early 1800s. For 150 years, Rome was the jewel in the crown of this artistic greatest-hits circuit. During the exact same time span, however, the political power of the papacy, and thus of the city, diminished dramatically. The same 1648 treaty that brought peace to Europe had drained much of the pope's influence in world affairs. Rome's relevance was now that of a cultural capital: an urban tableau brimming with spectacular set pieces.

Rome as Theater

The ambitious seventeenth-century publisher Giovanni Giacomo de Rossi was one of the first to capitalize on Grand Tourists' burgeoning appetite for graphic material relating to Rome. In collaboration with his protégé, the gifted etcher Giovanni Battista Falda, de Rossi issued a series of lavish illustrated books and eye-catching maps designed expressly to appeal to this market. The *Nuova pianta et alzata della città di Roma* (*New Plan and Elevation of the City of Rome*; fig. 102) of 1676 was their crowning achievement: the most complete and accurate visual record of Baroque Rome—the "world's theater"—offering a vivid glimpse of the great spectacle the city had become over the course of the 1600s. Rome's cityscape was a place of Church pageantry, festivals, and processions that amazed locals and foreigners alike. While Falda's map is devoid of human actors, it presents a glorious stage indeed.

A mural-scale etching printed from twelve copper plates on as many sheets of paper, the map stretches to approximately five by five feet. It takes the form of a bird's-eye view from high above Rome on a crystal-clear day, boasting a kind of hyperrealism where every feature of the city comes into perfect focus. The city's borders are defined, as usual, by the third-century Aurelian Wall, with the northernmost city gate, the Porta del Popolo, at left, St. Peter's and the Vatican below in the lower left corner, the Capitoline Hill roughly at center, and the Tiber running its curving course from left to right (north to south).

Signature Baroque urban projects appear in all their glory, particularly those undertaken in recent decades by Pope Alexander VII. His most prominent intervention was the transformation of St. Peter's Square (completed in 1667) into one of the largest, most scenographic piazzas in all of Italy (fig. 103). There, Gianlorenzo Bernini's curved double-colonnade funneling visitors toward the basilica was likened to the arms of the Church gathering the faithful in its embrace.

Additional building projects that Falda depicts resulted from private initiative. Opulent villas are scattered across the city's greenbelt at the upper half of

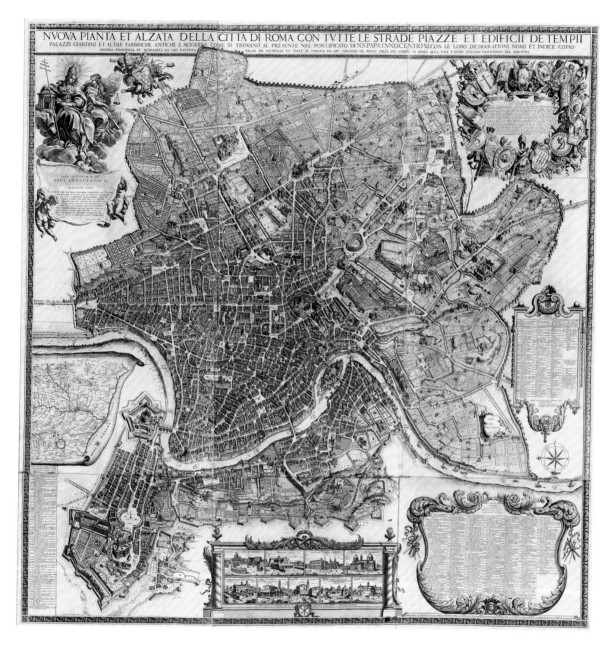

the image. These suburban retreats of wealthy, powerful families like the Barberini, Medici, and Montalto were rising stars in the constellation of Rome's cultural life. They were the locations of private collections and increasingly of cultural salons, where the literati eagerly gathered.

The cityscape is not all that grants the *Nuova pianta* its dramatic flair, for its edges are adorned with elaborate ornamentation. At the lower margin of the map is a cartouche containing small architectural views (or *vedutine*) of the major pilgrimage churches. At upper left, two female allegorical figures symbolize Justice and the Church—the latter, recognizable by her papal tiara, gazing down protectively at the city. Additional embellishments include lengthy

Fig. 102

Giovanni Battista Falda, *Nuova pianta et alzata della città di Roma*, engraving and etching, Rome: Giovanni Giacomo de Rossi, 1676. Vincent J. Buonanno Collection.

Rome of the Grand Tourists

Fig. 103

Falda, *Nuova pianta*, detail
showing St. Peter's

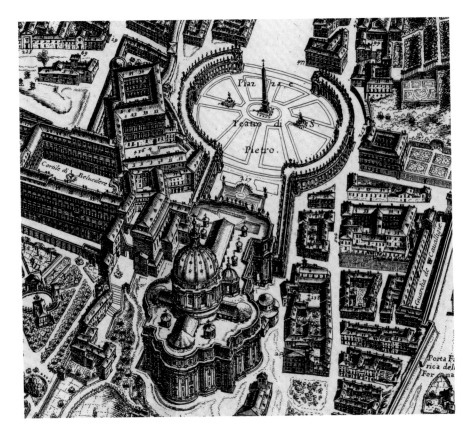

indices of Rome's important churches and palaces and an inset map of the
city's surroundings. At upper right, encircled by heraldry of the city's fourteen
administrative districts (or *rioni*), de Rossi included a public address inviting
"the noble and studious reader" to immerse himself in the map, "wandering
with the eyes through all the streets, piazzas, gardens, and quarters of the city."
His words eloquently describe the experience distant viewers hoped to regain
by gazing at a map like the *Nuova pianta*, which allowed them to imagine themselves back in the Roman cityscape, enjoying all the delights the city had to offer.

The Origins of the Tourist Plan

Although the popular guidebook industry and the accompanying tourist plan
tradition would not fully materialize until well into the 1800s, their roots can be
traced much earlier. In 1706, in particular, we find the first explicit statement
of a map being "very useful for travelers." Those very words appear at the top
of a plan published by François Nodot as part of his *Nouveaux mémoires de mr.
Nodot; ou Observations qu'il a faites pendant son voyage d'Italie* (*New Memoirs of
Mr. Nodot; or Observations He Made During His Voyage to Italy*). The book is a
hybrid: part travelogue, part instructive guidebook. It consists of two volumes,
the first of which addresses Rome's Christian sites, the second the city's secular monuments, ancient and modern. The personal element of the so-called

memoir is played down in favor of suggested site visits: a shift in tone that heralds later developments.

In another shift toward practicality, several maps graced the roughly five-hundred pages of Nodot's *Mémoires*, but the *Nouveau plan* illustrated here (fig. 104) was by far the most elaborate and informative. A large item, it was folded in right after the preface in the first volume—thus prominently placed at the beginning of the book to provide a spatial overview of the city. Nodot's map closely followed a 1667 prototype by Flemish etcher Lieven Cruyl, which depicted the city in the same oblong form cut off from its surroundings—a form ultimately borrowed from Tempesta's influential plan of 1593, which had been recently been reissued for a fifth time in 1662. Nodot, like Cruyl before him, emulates Tempesta's prototype by copying the earlier map's contours, streets, and orientation, as well as overall format. The city's foreshortened form suggests we are

Fig. 104

François Nodot, *Nouveau plan de la ville de Rome . . . tres utille pour les voiageurs,* engraving and etching, Amsterdam, 1706. National Library of Poland, via europeana.eu

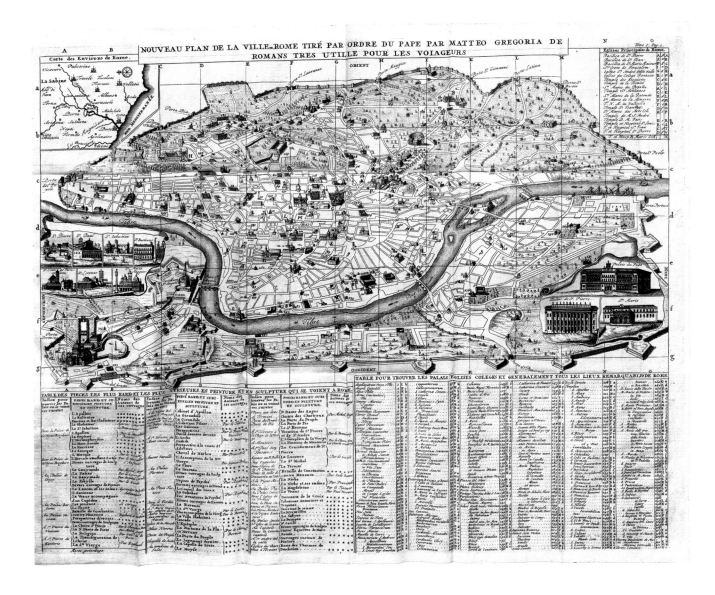

glimpsing it from above and to the west, and the monuments are also shown from a similar vantage point.

In Nodot's plan, as in Tempesta's and so many others, north is at left. Rome's perimeter is defined by its walls—the third-century Aurelian circuit in the background, the fortified seventeenth-century extensions on the crest of the Janiculum Hill in the foreground. Within the walls, Rome's "main street," Via del Corso, runs horizontally across the city from the Porta del Popolo at left. Beneath that thoroughfare, the Tiber wanders through the city from left to right, extending off the edge of the page at either side: the only element that would seem to link the city to the world beyond its borders.

Strikingly, Nodot, following Cruyl, has wiped away the multitude of ordinary structures from the Roman cityscape, leaving a select few highlights to depict as little pictures rising from the otherwise flat cityscape (fig. 105). At bottom left, for example, St. Peter's and the Vatican are depicted as three-dimensional "pop-ups." The map is part bird's-eye view, part accurate plan, helpfully spotlighting key monuments and the pathways connecting them. In this way it is very much like a modern tourist plan.

Perhaps the most touristy element, however, is the index at the bottom identifying the noteworthy places as correlated with letters, numbers, or zones on the map itself. Nodot lists the principal churches, palaces, and "generally all the remarkable places" of the city, as well as "the most rare and curious works of painting and sculpture that are to be seen in Rome." The map is a compendium of useful information for the intrepid cultural sightseer. It seems unlikely that it and the accompanying volume were carried around the city and used for on-site wayfinding: neither the book's organization, nor its heft, nor the map's

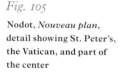

Fig. 105

Nodot, *Nouveau plan*, detail showing St. Peter's, the Vatican, and part of the center

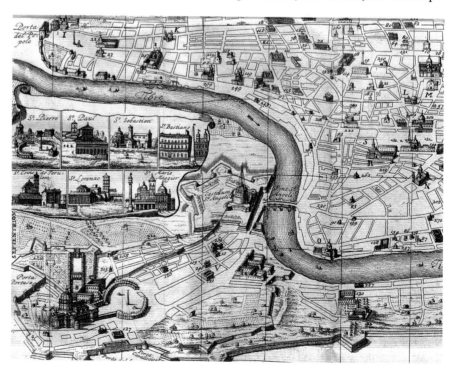

Chapter Seven

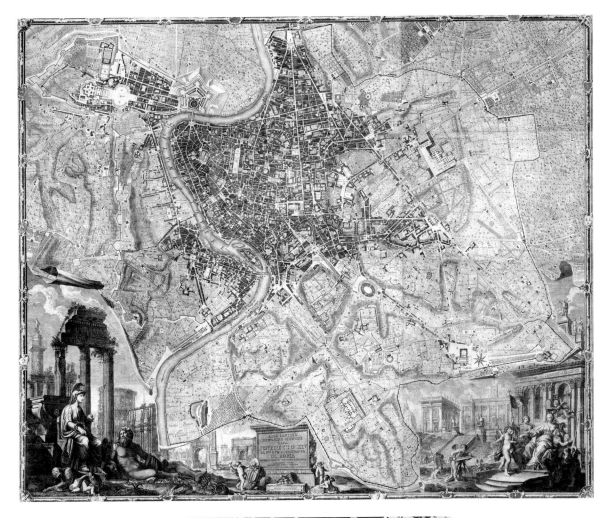

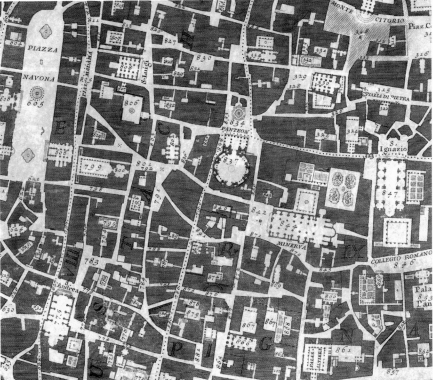

Fig. 106

Above: Giambattista Nolli, *Nuova pianta di Roma data in luce da Giambattista Nolli l'anno MDCCXLVIII*, engraving and etching, Rome, 1748. David Rumsey Map Collection, www. davidrumsey.com.

Fig. 107

Below: Nolli, *Nuova pianta*, detail showing the center (with the Pantheon at center and Piazza Navona at left)

Fig. 108

Nolli, *Nuova pianta*, detail
showing the decoration at
lower left

dimensions lend themselves to that scenario. But it is entirely conceivable that
visitors to Rome might have kept it in their lodgings for daily consultation.

The functionality of the map is complemented by a hint of the grand commemorative tradition that characterizes Falda's plan. Nodot, for example, includes small inset views of the seven main pilgrimage churches at the left, as

Chapter Seven

well as a few *vedutine* of prominent palaces at right. He was clearly aiming for a large, aesthetically pleasing map. Unlike the other maps discussed in this chapter, however, Nodot's was not meant as a showpiece. Subordinate to a book, it was intended primarily to orient the reader, whether that reader was using it to navigate city streets or to understand Rome from afar. If maps were not regularly used as travel aids in Nodot's time, his map hints that that day was coming.

Fig. 109

Nolli, *Nuova pianta*, detail showing the decoration at lower right

Rome Surveyed

With Giambattista Nolli's *Nuova pianta di Roma* (*New Plan of Rome*; fig. 106) of 1748, we move back into the realm of that grand commemorative tradition, but with a major difference. Comparable to Falda's plan in sheer scale and

Fig. 110

Giuseppe Vasi, *Prospetto dell'alma città di Roma*, etching, Rome, 1765. Vincent J. Buonanno Collection.

technical refinement, Nolli's image is revolutionary as a precisely measured ground plan, not a plausible view from above. Its form owes much to the ethos of the Enlightenment, when intellectuals placed their greatest faith in science and reason. Nolli was an engineer from Como, an expert surveyor at the head of a large team whose project to map Rome, initiated in 1736, was championed by influential members of the city's cultural and scholarly elite. More than a decade in the making, the *Nuova pianta* is a cartographic milestone, still celebrated for its accuracy rivaling that of modern digital maps.

Nolli's plan was printed on twelve sheets with an accompanying index to more than 1,300 numbered sites. With the sheets joined together, it stretched to almost six by seven feet. In a deliberate departure from previous maps of Rome, Nolli placed north at the top rather than left: an orientation that immediately

Chapter Seven

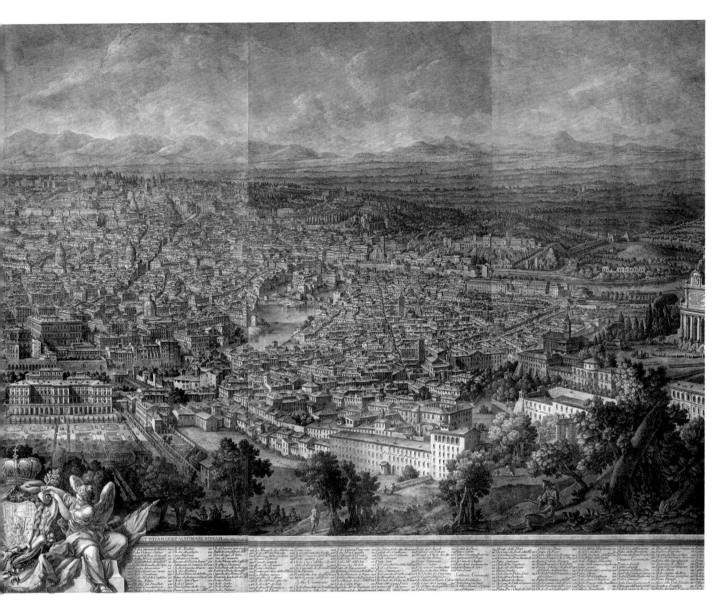

became standard. The Piazza del Popolo now crowns the image, the three Renaissance avenues that originate there radiating out into the city below.

Nolli depicted the urban fabric in an early variant of a form known to modern architects as a "figure-ground" plan. Public spaces like streets, piazzas, church interiors, and courtyards are left uninked—their contours defined by adjoining structures and walls—signifying accessibility, while private or closed off places are shown gray, denoting impenetrability (fig. 107). Nolli's thoroughness surpassed even Falda's, and his graphic conventions were unprecedented in their sophistication. He recorded the locations of drains and water fountains, took pains to distinguish buildings of different eras (showing ancient structures in solid black), and signaled hypothetical reconstructions by using outline only.

Rome of the Grand Tourists

Fig. 111

Vasi, *Prospetto*, detail
showing artist sketching

Nolli's map set a new benchmark for the scientific mapping of Rome and of all cities, but it was still geared primarily toward wealthy collectors. Its commercial, glorifying purpose is evident in its dazzling technique and decoration. The lower border of the map, designed by none other than Panini, abandons Enlightenment reason for an imagined Rome, more vibrant than the measured version above. In a clever trompe l'oeil flourish, the mapped city appears to curl upward at its southern edges, like irregularly cut paper, revealing the scene that unfolds beneath (fig. 108).

There, against a backdrop of whimsically arranged, famous ruins, a female personification of Ancient Rome sits enthroned amid an assortment of recognizable, if damaged, ancient statues. Posed stiffly, in shadow, she looks glumly to the right, where a more vivacious, brightly lit female figure with a decidedly more festive entourage returns her gaze. Framed by recent landmarks and surmounted by a hovering putto who crowns her with the papal tiara, this young beauty represents the Church, as well as New Rome (fig. 109). Together, these twins embody the city's most important identities—at least as far as Grand Tourists were concerned.

But perhaps the most revealing vignette in this artistic tour de force is the group of figures busily going about their business at the feet of New Rome. Here, four cherubs reenact the task of surveying the city: three use a folding chain to measure distances, while their chief—a surrogate for Nolli himself—records bearings on an advanced version of an instrument known as a plane table. These charming, playful characters are therefore engaged in a most serious mission: mapping the modern city.

A Panoramic Vision

An allegorical self-portrait also lies hidden in Giuseppe Vasi's sweeping view of Rome from 1765 (fig. 110): an artist half-submerged in the shadows of the foreground, his back to us as he sketches the city that stretches before his eyes (fig. 111). This figure gazes out at Rome from the Janiculum, the highest hill within city limits, marking Rome's western edge. Unlike Nolli's putto-surveyor, who measures the city using instruments, Vasi's artist draws Rome directly, as he sees it with his own eyes. The panorama he records was a favorite stop for Grand Tourists, the most popular "scenic viewpoint" from which to take in the city. Tempesta and others had also portrayed Rome from this vantage, but no previous artist had so vividly brought it to life before the eyes of spectators.

A native of Sicily, Vasi had arrived in Rome in 1736, where he set up his own print shop specializing in etched views catering to Grand Tourists' now insatiable desire for imagery of the Eternal City. His *Prospetto* of Rome was by far his most ambitious work. Printed on twelve sheets and measuring almost four feet high by over eight and a half feet wide, it includes an index at the bottom keyed to 390 Roman sites, organized into eight daylong itineraries. The image, however, is clearly not meant for plotting a walking tour: a viewer would search it in vain for a clear footpath.

Fig. 112

Vasi, *Prospetto*, detail showing St. Peter's and Castel Sant'Angelo

To be sure, it is possible to identify a number of highlights. St. Peter's—only slightly exaggerated in proportions—is squeezed in at the left even though it is not actually visible from the main crest of the Janiculum (fig. 112). To the right of it is the Castel Sant'Angelo, smoke billowing from its cannon. Across the meandering Tiber, careful inspection of the city center reveals the low dome of the Pantheon, among other landmarks. But overall, Vasi's atmospheric *Prospetto* recreates the experience of beholding the city as a whole in all its grandeur and beauty. Rome itself has become a work of art to be admired from afar.

Vasi employs some artistic license—the view seems taken with a wide-angle lens (or assembled piecemeal, somewhat like a modern, stitched-together photographic panorama), and from slightly above, not on, the Janiculum crest—but overall the *Prospetto* is scrupulously faithful to reality and unusually evocative. Spectators can almost feel the light breeze stirring the leaves in the foreground and the warm Roman sun on their face as they gaze out toward the city shimmering in the distance. More than Nolli's aloof, scientific cartography, Vasi's panorama transported Grand Tourists back to the sensory delights of their Roman sojourn.

FURTHER READING

Benson, Sarah. "Reproduction, Fragmentation, and Collection: Rome and the Origin of Souvenirs." In *Architecture and Tourism: Perception, Performance and Place*, ed. D. Medina Lasansky and Brian McLaren, 15–36. Oxford: Berg, 2004.

Bevilacqua, Mario. *Roma nel secolo dei lumi: Architettura, erudizione, scienza nella pianta di G. B. Nolli "celebre geometra."* Naples: Electa, 1998.

Black, Jeremy. *Italy and the Grand Tour.* New Haven: Yale University Press, 2003.

Haskell, Francis, and Nicholas Penny. *Taste and the Antique: The Lure of Classical Sculpture, 1500–1900.* New Haven: Yale University Press, 1981.

Hibbert, Christopher. *The Grand Tour.* London: Spring Books, 1969.

Krautheimer, Richard. *The Rome of Alexander VII, 1655–67.* Princeton: Princeton University Press, 1985.

Maier, Jessica. "Giuseppe Vasi's *Nuova Pianta di Roma* (1781): Cartography, Prints, and Power in *Settecento* Rome." *Eighteenth-Century Studies* 46 (2013): 259–79.

McPhee, Sarah. "Falda's Map as a Work of Art." *Art Bulletin* 101 (2019): 7–28.

Sánchez-Jáuregui Alpañés, María Dolores, and Scott Wilcox, eds. *The English Prize: The Capture of the* Westmorland, *an Episode of the Grand Tour.* New Haven: Yale University Press, 2012.

Tice, James T. "'Tutto insieme': Giovanni Battista Falda's Nuova Pianta di Roma of 1676." In *Piante di Roma dal Rinascimento ai catasti*, ed. Mario Bevilacqua and Marcello Fagiolo, 244–59. Rome: Artemide, 2012.

Tice, James T., and James G. Harper. *Giuseppe Vasi's Rome: Lasting Impressions from the Age of the Grand Tour.* Eugene, OR: Jordan Schnitzer Museum of Art, 2010.

Wilton, Andrew, and Ilaria Bignamini, eds. *The Grand Tour: The Lure of Italy in the Eighteenth Century.* Exhibition catalogue. London: Tate Gallery Publishing, 1996.

Chapter Seven

Chapter Eight

Rome of the Mass Tourists

"I have already met three 'flocks' [of tourists] and anything so uncouth I never
saw before. . . . The cities of Italy are deluged with droves of these creatures,
for they never separate, and you see them, 40 in number, pouring along a
street with their director . . . circling around them like a sheep dog and really
the process is as like herding as may be."

Cornelius O'Dowd (a.k.a. Charles Lever)

Although he was a fictional character created by Irish novelist Charles Lever,
the words of pseudo-essayist Cornelius O'Dowd disparaging his compatri-
ots from the British Isles as they overran Italy probably ring familiar to anyone
who has spent time in Rome, Florence, or Venice in the present day. Whether
you have taken part in a large group tour, disembarked from a mammoth cruise
ship in the Venetian lagoon that really has no business being there, or plotted
your own bespoke journey through Italy's most popular places, chances are
you have encountered just such a crowd of sightseers, if not counted yourself
among them. These uncouth creatures—as a snob like Lever would have it—
are a distinctly modern phenomenon, dating to the 1800s, specifically to after
1815, when Napoleon was defeated and travel between the British Isles and Italy
once again became viable.

At that moment, the floodgates opened, and Rome became a sought-after
destination to a new class of visitors: the hordes of nonelite tourists who came
from all over Europe to experience the city's cultural riches. Lord Byron, who

Fig. 113

"English Tourist," from
World's Dudes series for
Allen & Ginter Cigarettes,
1888. Metropolitan Museum
of Art, The Jefferson R.
Burdick Collection, Gift
of Jefferson R. Burdick.

traveled through Italy in high Grand Tour fashion in 1817, believed that their influx was temporary. Rome, he wrote, "is pestilent with English. . . . A man is a fool now who travels in France or Italy, till this tribe of wretches is swept home again." "In two or three years," he continued, "the first rush will be over, and the Continent will be roomy and agreeable." Byron was wrong: the trend was on the rise, never to be reversed, except relatively briefly in times of war.

Differentiated from Grand Tourists by their sheer numbers, more modest means, and less than luxurious travel practices, this new group heralded the beginnings of popular tourism. Carrying suitcases, they arrived in second-class railway cars, lodged in pensions or hostels, and ate in taverns. Rather than hiring professional guides, many signed up for budget-friendly package tours, while others came armed with a do-it-yourself attitude and a new kind of publication, the guidebook (fig. 113).

Increasingly, they relied on mass-produced tourist plans that signaled points of interest as well as services geared toward daily needs—not only monuments but also hotels and dining establishments together with the routes linking them, and eventually public transport lines. This lowly and practical category of cartography, the subject of this chapter, presents a stark contrast to the splendid maps discussed in the previous one. The items range from folding plans included in popular guidebooks issued by Karl Baedeker and others to stand-alone plans like one by Romolo Bulla, geared toward quick, convenient consultation and portability. Generally speaking, production values plummeted while utility surged.

Beyond the maps and related tools they used, mass tourists were also set apart by the speed with which they took in the sites, often compressing into weeks what the Grand Tourists spaced out over a year or two. As is typical for vacationers today, these humbler visitors couldn't afford to stretch their trips over an extended period: they had responsibilities at home, a living to make. By design, their expeditions were brief and superficial holidays, not profound, life-changing experiences. While Grand Tourists allotted Rome a minimum of six weeks and often considerably longer, travelers who signed up for Thomas Cook's whirlwind tour of Italy might spend two and a half days in the city. They were seeing the sights, not immersing themselves in local history and culture: travel, for them, was about recreation more than education.

A number of factors converged to fuel the beginnings of leisure travel in the mid- to late 1800s. The industrial revolution had led to growing prosperity in general and specifically to an expanded upper-middle class, which had some disposable income. In a sense, their expenditures on travel were aspirational: they were emulating the travel practices of their wealthier brethren, albeit in abbreviated form and on a budget. There is no doubt that many tourists were motivated by a genuine passion for seeing new places, but it is also true that tourism became a kind of conspicuous consumption.

Tourists still came primarily from Britain, to a lesser extent Germany and other northern European countries, and still less from the United States. A crucial factor that enabled them to make long-haul journeys was the considerable improvement in transportation technologies and networks. In the 1700s, Grand Tourists seeking to avoid passing over the Alps by opting to go by boat from, say, Nice to Genoa risked seasickness and frequent delays due to inclement weather. The most common craft was a crude vessel known as a felucca, which came equipped with sails and oars but was not fit for stormy or windy conditions. Sea travel became considerably less complicated with the advent of the steamship. The first one crossed the English Channel in 1816, and others were soon navigating the larger rivers of continental Europe. Regular service between Marseilles and ports of western Italy began in the 1830s.

For those traveling by land, Napoleon had improved or built new roads throughout much of Italy, France, and Switzerland, which simplified passage over the Alps and facilitated movement from one place to another generally. Many tourists could make the journey in large public stagecoaches known as diligences—the predecessors of the modern Greyhound bus, although slower and bumpier. Alternatively, people could travel those same routes by "post" (hiring a carriage and horses for personal use, to take from one point to another, usually coinciding with spots that offered lodging, meals, and fresh horses to use for the next leg of the voyage). Within Italy, there were *vetturini*—private guides who accompanied small groups of tourists to their destination, arranging transport, accommodations, and meals along the way. This mode of travel was popular because it offered "all in one" services at a low cost, but it was also notoriously slow and uncomfortable.

All this changed for good once the era of rail travel began in the 1840s. Although the railroad systems in Italy and Switzerland were slower to develop than in Britain and France, by 1860, it was possible to travel from London to Naples by train in a matter of days, whereas just five decades earlier the same journey by other means had taken months. Then, following Italian unification in 1861, the peninsula's rail system expanded further—even if it was composed of multiple companies until the rail system was fully nationalized, in 1905, with the creation of the Ferrovie dello Stato. In contrast to Grand Tourists of the 1700s, who in spite of their means had made their way southward slowly and arduously, by private carriage or even litter (the latter for crossing the Alps), often with the fear of bandits lurking along the path, lesser tourists of the 1800s could ride in relative comfort by sea or by rail while spending a lot less time, money, and effort to reach their destination.

Improvements in transportation, in turn, helped to usher in the origins of budget travel. Thomas Cook, the first true impresario of tourism, stands out as the key figure who harnessed new means of transport like the railroad to convey people across distances quickly and inexpensively. By inventing now-ubiquitous strategies such as group rates in order to arrange economical and convenient journeys, he smoothed out the bureaucratic and logistical challenges of travel that had previously put it out of reach for anyone with limited time or funds.

Cook invented the package tour in the 1840s, leading groups first to places in the British Isles—places that now sound distinctly unexotic, like Liverpool—and then to Switzerland, Italy, and even Egypt by the 1860s (fig. 114). Wherever they went, his clients traveled at a dizzying pace, cramming in as much as possible in a short time. Cook himself was a fascinating figure: a onetime Baptist missionary and temperance campaigner turned entrepreneur, he relished his role in democratizing travel. In response to elitists like Lever and Byron, he wrote, "God's earth, with all its fullness and beauty, is for the people; and railroads and steamboats are the results of the common light of science, and are for the people also."

Motivated by more than commercial impulses, Cook saw his role in political terms, viewing his tours as a way to help the cause of uniting Italy by treating the peninsula's diverse parts as part of a single overarching culture. The origins of popular tourism, in fact, coincide perfectly with the period known as the Risorgimento, the decades-long nationalist movement riddled with republican uprisings and revolutions, as well as repressive countermeasures. The Risorgimento concluded with the official unification of Italy as a single nation-state only in 1861 and the establishment of Rome as the country's capital a decade later. United in theory, in practice the country was stitched together from profoundly (perhaps irreparably) discordant cultural and linguistic entities, which to this day sit uneasily together. Cook, optimistically, aimed to spread the perception of Italy as a unified nation to foreign visitors and natives alike, while also bringing much-needed tourism revenue to a country in crisis.

At the same time that Cook was seeking to open the possibility of travel to the many, great works of art that had once been off limits to all but the few were becoming accessible. Some public collections had begun opening their doors in the seventeenth century, but the origins of many of today's most prominent art museums date from the mid- to late 1700s, when they took their place in the larger realm of public collections relating to botany, zoology, natural history, and ethnography. The British Museum, which included drawings and paintings among other more encyclopedic offerings, opened in 1759, the Uffizi Gallery a decade later, the Prado (as it was later named) in 1785.

This phenomenon was deeply rooted in the Enlightenment—the age of reason. Organized displays of artworks, like those of fossils, plants, bones, minerals, and all sorts of specimens and exempla, were one way to make sense of the world in all its diversity and of its evolution over time—of categorizing branches of knowledge within a variety of frameworks (geographical and chronological, not to mention national, religious, racial, ethnic, and so on). At the same time, works of art were understood as highlights of human achievement, distinguished by their aesthetic value, so art was recognized as something different and special: a mark of civilization. Some recognized masterpieces became destinations in and of themselves as they began to rise toward icon status in the Western canon.

Often the birth of a museum was deeply tied to the turbulent political climate. The definitive opening of the Louvre, for example, came after French Revolutionaries ousted the monarchy and, in a symbolic gesture, made publicly available great works of art that had previously been shut away in the royal palace. In Rome, the papacy moved to make many of its magnificent art collections available to the populace as a sign of beneficence—a public-relations move calculated, in part, to avoid sharing the fate of the French royalty. The Capitoline Museum, which had long been semi-accessible, was opened as a fully operational museum in 1734. Sections of the Vatican began to follow in the 1770s, a process that continued into the 1800s.

The tourism industry was also dependent on the rise of mass-produced guidebooks. The surge in travel after 1815 fueled this genre, in which the flowery language of Grand Tour travelogues and itineraries—those prolonged, idiosyncratic meditations on art, beauty, and life—were left behind in favor of practical advice dispensed in an objective, matter-of-fact tone. These books embraced a distinctly prosaic style with no hint of the poetic pretensions that often characterized the memoir. The most important publishers were John Murray in England and Baedeker in Germany, who established a format and a series of "must see" spots that have changed remarkably little in the past century and a half (fig. 115). Originally the two guidebooks were very similar, with Baedeker copying Murray and eventually unseating him after he began publishing in English.

In World War I, the English editors of Baedeker's guides acquired the rights to Murray's competing handbooks and started a new line called the Blue Guides, which are still in print and a favorite of more scholarly inclined travelers. Nowadays there are guidebooks aimed at a wide variety of tourists, budgets, and interests. From Fodor's and Frommer's to Rick Steves, Lonely Planet and Rough Guide to DK Eyewitness, there are countless variations, but all of them are fundamentally indebted to Murray and Baedeker. Even in our increasingly digital age, no app has yet appeared—as of this writing—that has come close to challenging the physical guidebook for convenience, functionality, and popularity.

Most modern guidebooks offer some kind of capsule history of Italy to the present day. For the most part, nineteenth-century versions barely acknowledged the current political situation, even though these years were some of the most turbulent in Italian history. There is no place in Baedeker's and

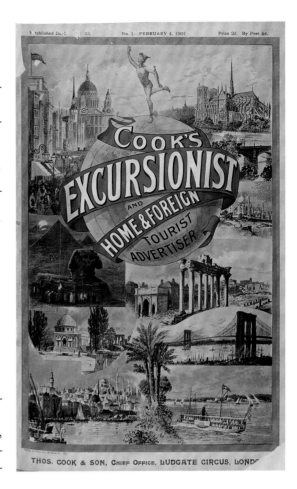

Fig. 114

Cover of *Cook's Excursionist*, 1902. Thomas Cook Archives.

Rome of the Mass Tourists

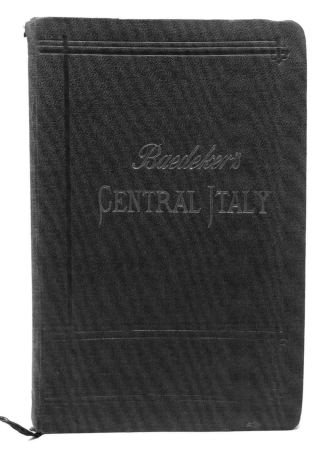

Murray's volumes for the violence and upheaval of mid-nineteenth-century Italy. As packaged for foreign tourists by these entrepreneurs, the country was all about the monuments of the past, not the Italian people living in the troubled present.

The maps discussed in this chapter all emerge from this fraught time, with its paradoxical combination of popular (foreign) tourism and popular (native) unrest. Like the tourists whose needs they reflected, these maps were not only more modest than their seventeenth-century predecessors but also less symbolic or allegorical. In Rome, the golden era of magnificent display maps peaked in the 1600s and 1700s, then gradually faded—even as maps became commonplace travel tools. Not coincidentally, this development parallels Rome's shift from political center to tourist mecca. After the cartographic glories and vast dimensions of works described in previous chapters—by Tempesta and Maggi, as well as Falda, Nolli, and Vasi—few publishers continued producing grand maps or bird's-eye views to embody the glories of Rome. But the tourism industry continued, and continues, to grow exponentially—along with the rich variety of maps that cater to it.

Fig. 115

Cover of *Baedeker's Handbook to Central Italy and Rome*, 1872. Emory University Digital Library, via archive.org.

The Guidebook Impresario's Rome

It would be hard to overstate the importance of Karl Baedeker in the history of the modern guidebook. Baedeker's line of English-language guidebooks began to come out in the 1860s, with the first being his *Rhine* of 1861. His aim, as stated in the preface to his 1872 *Handbook for Travellers to Central Italy and Rome*, was "to render the traveller as independent as possible of the services of guides, valets-de-place, and others of the same class, to supply him with a few remarks on the progress of civilisation and art among the people with whom he is about to become acquainted, and to enable him to realise to the fullest extent the enjoyment and instruction of which Italy is so fruitful a source."

In other words, Baedeker was determined that his guidebook would free travelers from dependence on any assistance outside of, well, his guides. In this way, he effectively replaced the *cicerone* who had been such a fixture for Grand Tourists, even if he could not replace the cheaper sort of human "herder" mocked by Charles Lever—namely, the group leader of any Cook's tour. Regardless, it was a winning proposition, as proven by the astonishing success of Baedeker's bestselling line of guidebooks.

More than Murray, Baedeker also emphasized that his volumes were geared toward budget travelers, thereby explicitly catering them to a middle-class demographic. He wanted, he wrote in the *Handbook*'s preface, "to place the traveller in a position to visit the places and objects most deserving of notice with the greatest possible economy of time, money, and, it may be added, temper; for in no country is the traveller's patience more severely put to the test than in some parts of Italy." Baedeker expanded on the last statement later in the guide's lengthy introduction, in a section titled "Intercourse with Natives," where he painted a picture of Italians as impetuous, cunning opportunists, always on the lookout for a sucker. "The traveller," he warned, "is regarded by landlords, waiters, drivers, porters, and others of the same class, as their natural and legitimate prey."

Even as Baedeker perpetuated what was to become a longstanding cultural stereotype about the perils of dealing with greedy, sneaky Italians, he also made good on his promise to grant independence to tourists by providing an unprecedented amount of information about travel, routes, destinations, lodging, dining, landmarks, and so on. He was also the first to devise a system of rating sites

Fig. 116

Eduard Wagner, folding map of Rome from Karl Baedeker, *Handbook for Travellers to Central Italy and Rome*, Coblenz, 1872. Image joined digitally by author. HathiTrust/University of California, Los Angeles, via archive.org.

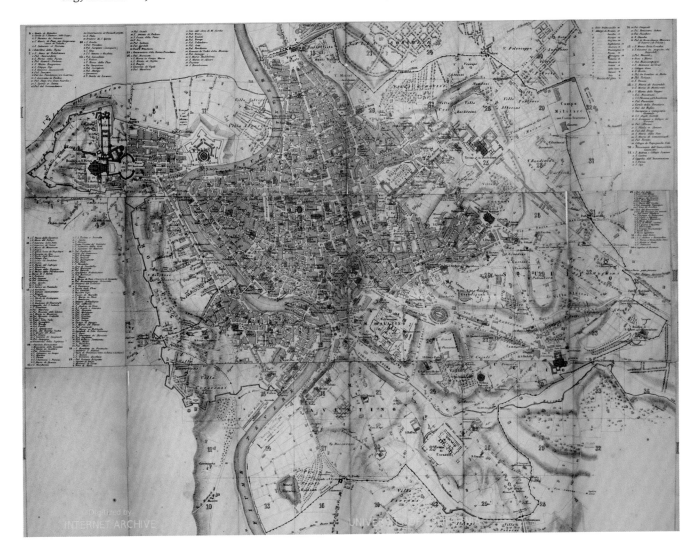

by using stars—a system that has become ubiquitous today. Baedeker realized that maps were essential to making travelers self-sufficient, and he hired skilled cartographers to produce sophisticated plans for his guidebooks. Maps have played a key role in guidebooks ever since.

From its first edition in 1867, Baedeker's *Handbook for Travellers to Central Italy and Rome* included many plans. For Rome alone, there was the large folding plan illustrated here (fig. 116), which was tucked inside the back cover, a smaller key map correlating to written sections of the guide, a map of ancient Rome, and a detail map of the Forum. The ability for readers to cross-reference cartographic and written information, and to choose their own paths between and across the guide (and the city itself), will ring very familiar to modern guidebook users.

The large plan of Rome from Baedeker's guidebook was based on an up-to-date, scientific Italian prototype and printed in Darmstadt, Germany, by Eduard Wagner. In size, form, and relative accuracy it is comparable to seventeenth-century luxury views, particularly that by Falda discussed in the previous chapter (fig. 102). There, however, the similarities end. Baedeker's map is a straightforward and detailed street plan with no embellishment to cloud its functionality. Meant to be folded into the small contours of the guidebook for portability, it was printed on thin, inexpensive paper using lithography—a medium better suited for mass production than the older methods of etching or engraving.

The map was also divided into three sections in order to obviate "the disagreeable necessity of unfolding a large sheet of paper at every [on-site] consultation" (the version illustrated here has been joined digitally). Because the use of maps by tourists in this fashion was a relatively new phenomenon, Baedeker provides a helpful tip to novices: "The inexperienced are recommended, when steering their course with the aid of a plan, to mark with a coloured pencil, before starting, the point for which they are bound. This will enable them to avoid many a circuitous route."

On the map, one can easily imagine plotting a path (with or without colored pencil) within the clearly delineated street network while making one's way to one of the many churches, ancient monuments, palaces, and other sites listed in the three indices at the edges. Among other noteworthy recent features, the map shows the provisional location of Rome's Termini train station at upper right. A temporary station had been set up at this spot in the city's northeastern quadrant in 1863, to be replaced by a permanent version begun in 1868 and completed in 1874 (that station, in turn, was rebuilt in the twentieth century).

Informative as it is, however, the map is not inherently touristy. There is no attempt to signal the locations of amenities such as hotels and taverns, nor are publicly accessible sites privileged over private ones that would have been off limits (and thus irrelevant) to visitors. Such information was, however, readily available in the guidebook itself, with which the map was meant to be correlated. In this way, the guidebook as a whole, including its cartographic contents, is akin to a modern web interface that allows people to click on links to access information in personalized sequences, charting their own paths through the digital world.

Rome for a Rather Important Woman Traveler

This fairly banal map (fig. 117) has a remarkable provenance. First, a bit of background. It was included in a guidebook published in 1875 by the imposingly named Shakspere Wood, a decades-long British expat in Rome who wrote from considerable personal knowledge of the city. Wood's biography is a bit of a cliché: after training at London's Royal Academy as a sculptor, he went to Rome to "perfect himself in that branch of art" (as his 1886 obituary stated). The lure of the Eternal City proved too great to resist, and Wood stayed on permanently, finding work as a foreign correspondent for the London *Times*. His guidebook, published under the auspices of Thomas Cook's tour company, was meant to snatch a bit of the market cornered by Baedeker's *Handbook for Travellers*. With

Fig. 117

W. & A. K. Johnston, map of Rome, from Shakspere Wood, *The New Curiosum Urbis: A Guide to Ancient and Modern Rome* (London: Thomas Cook, 1875). Lincoln Financial Foundation Collection, via archive.org.

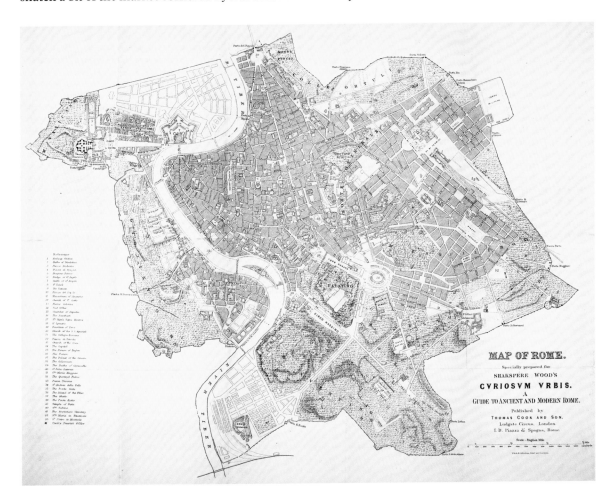

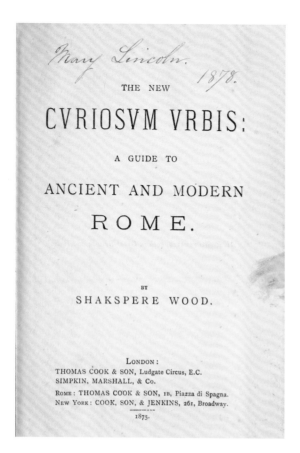

THE NEW

CVRIOSVM VRBIS:

A GUIDE TO

ANCIENT AND MODERN

ROME.

BY

SHAKSPERE WOOD.

LONDON:
THOMAS COOK & SON, Ludgate Circus, E.C.
SIMPKIN, MARSHALL, & Co.
ROME: THOMAS COOK & SON, 1B, Piazza di Spagna.
NEW YORK: COOK, SON, & JENKINS, 261, Broadway.

1875.

Fig. 118

Wood, *The New Curiosum Urbis*, detail showing title page with Mary Todd Lincoln's signature

only one edition appearing, it did not quite manage to do so, but that one version of 1875 seems to have sold decently.

The map of Rome that graced Wood's *Curiosum Urbis* was printed in Edinburgh by the cartographic firm of William and Alexander Keith Johnston. A folding map like Baedeker's, it is smaller and less detailed, with only forty-three sites indexed, together with the location of Cook's Tourist Office, centrally situated in Piazza di Spagna (a location that has a long history as a cultural and informational hub for foreign travelers). The map would have been perfectly adequate as a street plan and basic tool for orientation. In addition to the historic center, which had changed little in the past century, it shows new neighborhoods planned northeast of the Vatican (the area known today as Prati di Castello, or simply Prati) and at the southern end of the city (Testaccio). New quarters defined by grid plans also appear around Termini train station to the east. All of these developments reflect Rome's increased importance and expected surge in population following 1871, the year it became capital. Overall, Wood's map is a functional if unremarkable mass-produced item.

More notably, this particular copy of Wood's *Curiosum Urbis* was the property of Mary Todd Lincoln, whose signature graces the title page, along with the handwritten year of 1878 (fig. 118). After the assassination of her husband by John Wilkes Booth in 1865, as well as the death of her son Thomas (Tad) Lincoln in 1871 following the earlier deaths of two other sons, Mary Todd Lincoln endured bouts of crippling grief and depression. Throughout the early 1870s she was in a fragile state of mental health, at one point being committed to a mental institution. Not long after her release, in 1876 she departed for Europe, where she stayed for four years. During that time she explored parts of Italy, including Rome, while based in Pau, France.

Mary Todd Lincoln's biography would seem to present a particularly compelling backstory for a visit to Rome: a desire to escape from personal tragedy while immersing oneself in the broader sweep of history, where anyone's individual circumstances are reduced to insignificance. On a larger sociological level, Lincoln's story raises important points about just who was traveling in that first wave of popular tourism. First, many more Americans began traveling to Europe during this time, especially in the relatively prosperous years after the Civil War. Indeed, transatlantic tourism surged in the second half of the nineteenth century. In contrast to the case with European travelers, however, for Americans such journeys were still largely reserved for the wealthy. Middle-class Americans would not benefit from the Thomas Cook effect until the twentieth century.

Another issue raised by the case of Mary Todd Lincoln is gender and travel. The Grand Tour had been a male-dominated phenomenon, while Cook's tours appealed more to families. Women did not just travel in the company of their spouses and relations, however. The nineteenth century witnessed a boom in the number of women traveling alone, which became a more socially acceptable practice. The vast majority of guidebooks were written by and for men, but intrepid women were setting out on their own in ever-greater numbers. By the late 1870s, Mary Todd Lincoln would not have stood out wandering through the Forum, a guidebook tucked under her arm.

Rome in Your Pocket

Romolo Bulla issued this pocket-sized folding plan (fig. 119) as a stand-alone item, a *Pianta Guida*, or *Plan-Guide*, meant to offer its users a quick reference to the locations of Rome's major monuments, as well as the most efficient means of getting from point A to point B. A chromolithograph—a relatively new and inexpensive method of color printing in which each hue was printed from a

Fig. 119

Romolo Bulla, *Pianta guida della città di Roma*, chromolithograph, Rome, 1885. Harvard Map Collection.

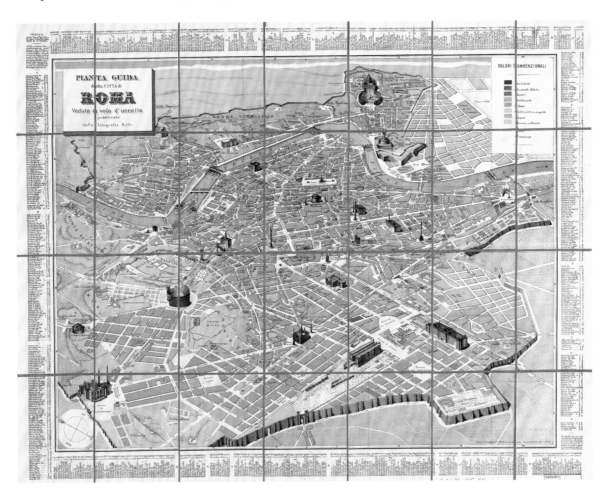

different stone—the map includes a color-coded key to identify ruins, hotels, theaters, churches, fountains, and other urban features of note for tourists. Around its edges, meanwhile, is a helpful index to Rome's streets.

Rome is shown in a bird's-eye view rather than a pure footprint, granting a bit of the illusion of seeing it from high above. The perspective is from the east, roughly above the position of Termini train station, which is conspicuously placed at the lower margin (fig. 120). If medieval maps like those discussed in chapter 3 often reflect the "pilgrim's perspective" of a visitor arriving by foot from the north, Bulla's view substitutes the "tourist's perspective" of the droves who now arrived by rail. By the late nineteenth century, Termini—not the Porta del Popolo—was the major gateway into the city.

But perhaps the clearest sign of the touristic purpose of Bulla's map is its approach to depicting Rome's urban fabric. Most city blocks are portrayed in abbreviated, flat plan, save for a small selection of the city's most important sites, which rise from the foreshortened city as disproportionately large pop-ups. As we have seen in previous chapters, this was a common strategy for signaling points of interest to a given target audience (see for example figs. 85, 94, and 104). It is the cartographic equivalent of Baedeker's star system, still often encountered in tourist plans today. In Bulla's view, just a handful of places are singled out for such treatment, including the Vatican and St. Peter's, the Colosseum, the Lateran Basilica, and the Castel Sant'Angelo.

The city was developing at a dizzying pace in the aftermath of 1871, and Bulla's map documents many recent changes. A new, major artery appears toward the center: Corso Vittorio Emanuele II, only recently completed (fig. 121). Running roughly along the route of the medieval Via Papalis, this east-west avenue was a continuation of Via del Plebiscito that originated at Piazza Venezia, leading in the direction of the Vatican on the other side of the Tiber.

Fig. 120

Bulla, *Pianta guida*, detail showing Termini and surroundings

Chapter Eight

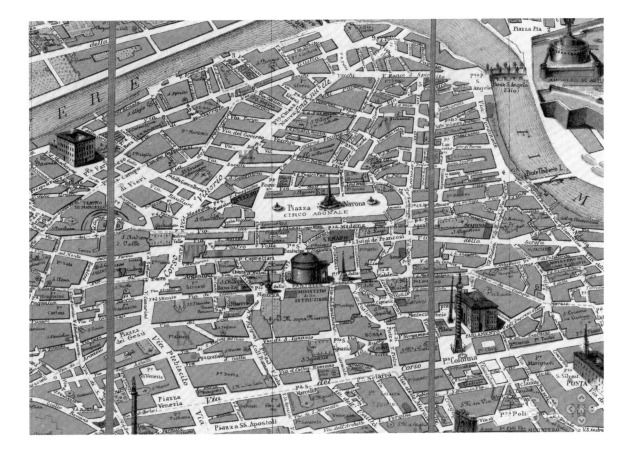

Fig. 121

Bulla, *Pianta guida*, detail
showing Corso Vittorio
Emanuele II

Rather than a new street per se, Corso Vittorio resulted from the expansion of existing streets to create a wide if irregular axis through Rome's center. In that way its creation was actually a relatively sensitive intervention, more so than many others that were to come.

Bulla also distinguished projected quarters of the city from existing parts, employing a relatively light taupe for the former vs. a darker terracotta hue for the latter. A number of new sections were in various stages of planning and construction in the 1870s and 1880s: not only Prati near the Vatican, but also the zones flanking both sides of Termini and around the new Piazza Vittorio Emanuele II on the Esquiline, Celio (the sloping neighborhood between the Colosseum and the church of Santo Stefano Rotondo), the area along the new Via Nazionale, the Aventine, and Testaccio. The telltale sign of a late nineteenth-century planned neighborhood is its grid plan with straight (or sometimes diagonal) streets delimiting ample city blocks. Of course, the Renaissance had had its share of straight streets, too, but they tended to be narrower and to frame smaller blocks, generating a more tightly knit urban fabric.

Another sign of modernization is Bulla's inclusion of Rome's new tram lines, which appear as dotted gray lines tracing a few select routes along the city's main thoroughfares. This modest beginning for Rome's mass transit system points forward to a future choked with buses, trams, and taxis (as well as the Metro if you count what goes on underground, as you always should in Rome).

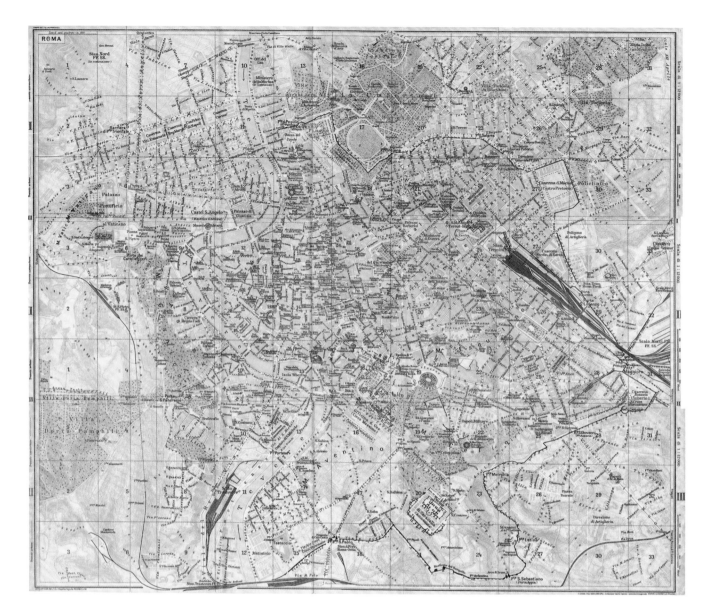

Fig. 122

Map of Rome from the
Touring Club Italiano's
Roma e dintorni (Milan,
1925). Private collection.

Naturally, there is no hint of that eventuality on Bulla's map, which is cleared
of human and vehicular traffic to make it as legible as possible for its intended audience.

Rome for Italian Tourists

Italian domestic tourism finally began to take flight toward the end of the nineteenth century. The Touring Club Italiano, which made its debut in 1894 as a
cycling association, was the first major entity dedicated to encouraging Italians
to see their country. Its founder, Luigi Vittorio Bertarelli, followed Thomas
Cook in seeing mass tourism as a way to help unite the patchwork that was Italy.
Within a couple of decades, the group had amassed more than 100,000 members as it moved beyond biking to focus on tourism more generally.

An early version of the TCI guide was published in 1895. Distributed to club members for free, it was geared primarily toward cycling itineraries. What was to become the organization's signature series of guidebooks, called the *Guide rosse* or *Red Guides* due to their characteristic red covers (a color scheme borrowed from Murray and Baedeker), first began to appear in 1914. The much-vaunted project was conceived in nationalist terms, as a way to free Italians from dependence on foreign guidebooks in order to acquaint them with their own homeland.

To this day, the TCI guides—which are still red, and cover all of Italy's regions in twenty-three exhaustive volumes—are considered the gold standard for scholarly guides to Italy. They furnish an almost overwhelming amount of authoritative information, often stretching to over one thousand pages. New editions come out frequently, scrupulously updated to incorporate the latest research and discoveries. The TCI guides have been renowned for their cartography since the early days of the organization, perhaps because the original focus on cycling meant that members needed better wayfinding tools than did, say, train travelers who did not need to navigate their own path. Each volume includes dozens of highly detailed, accurate maps and plans. The current edition of the Rome guide, in fact, contains more than one hundred.

This map of Rome (fig. 122) comes from the first edition of the TCI's *Roma e dintorni* (*Rome and Surroundings*), which was published to coincide with the Holy Year of 1925. This moment was one of transition in Rome, as the capital was poised on the edge of descent into ultranationalism. The "Roman Question" regarding sovereignty over the city had yet to be formally decided, but the papacy had finally seen the writing on the wall after decades of self-imposed, ineffective exile. Relations between the government and the Vatican had improved since the last Jubilee, in 1900. Now, a new administration was in power, namely the Fascist party led by Benito Mussolini, who saw a strategic advantage to fostering closer ties with the papacy. There was an atmosphere of reconciliation in the city. The Jubilee of 1925 was fairly subdued but still an occasion for cautious celebration. The TCI map of Rome, like the larger guide to which it was appended, gave no indication of the current, changeable, in retrospect ominous political situation—but such was the historical backdrop and motivation for its publication.

The map's overall cartographic quality and production values far transcend the other works discussed in this chapter. Included with the guidebook in its own booklet, the map was drawn at a scale of 1:12000. It is considerably more detailed than others we have seen—indeed minutely detailed—yet somehow also more legible. In part this graphic clarity is due to the precise color printing and coding, which allows for different categories of features to be understood intuitively as groups. A variety of fonts are also

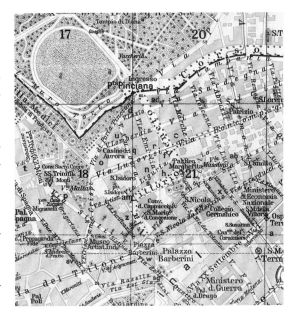

employed in correlation with separate categories of features and sites. The map includes a wealth of information: bus and tram lines, rail stations, ancient sites and buildings, as well as topographical differentiation and altitudes (many of Rome's larger hills, for example, are delineated with contour lines).

The map demonstrates that the city's development had continued apace since Bulla published his *Pianta Guida* four decades before. The area framing the swanky Via Veneto, on the northern side of the city leading toward the Pincian Hill, is now a reality (fig. 123). Within decades, it will be renowned as a magnet for glitterati the likes of Federico Fellini, Marcello Mastroianni, and Sofia Loren. But a lot was to transpire—to the city, to Italy, and to the world—before *La Dolce Vita* materialized in Rome.

FURTHER READING

Baker, Jean H. *Mary Todd Lincoln: A Biography.* New York: Norton, 2008.

Blennow, Anna, and Stefano Fogelberg Rota, eds. *Rome and the Guidebook Tradition: From the Middle Ages to the 20th Century.* Walter de Gruyter GmbH: Berlin, 2019.

Hart, Douglas. "Social Class and American Travel to Europe in the Late Nineteenth Century, with Special Attention to Great Britain." *Journal of Social History* 51, no. 2 (2017): 313–40.

Hom, Stephanie Malia. *The Beautiful Country: Tourism and the Impossible State of Destination Italy.* Toronto: University of Toronto Press, 2015.

Hooper-Greenhill, Eilean. *Museums and the Shaping of Knowledge.* London: Routledge, 1992.

Kallis, Aristotle A. *The Third Rome, 1922–1943: The Making of the Fascist Capital.* New York: Palgrave MacMillan, 2014.

Kostof, Spiro. *The Third Rome, 1870–1950: Traffic and Glory.* Berkeley: University Art Museum, 1973.

Palmowski, Jan. "Travels with Baedeker: The Guidebook and the Middle Classes in Victorian and Edwardian Britain." In *Histories of Leisure*, ed. Rudy Koshar, 105–30. Oxford: Berg, 2002.

Simmons, John E. *Museums: A History.* Lanham, MD: Rowman & Littlefield, 2016.

Withey, Lynne. *Grand Tours and Cook's Tours: A History of Leisure Travel, 1750–1915.* London: Aurum Press, 1997.

Chapter Nine

Rome Enters the Modern Age

"We must liberate all of ancient Rome from the mediocre construction that disfigures it, but side by side with the Rome of antiquity and Christianity we must also create the monumental Rome of the twentieth century. Rome cannot, must not, be solely a modern city . . . it must be a city worthy of its glory."

Benito Mussolini

Rome's existence is sometimes framed in terms of three distinct epochs: the First Rome of the emperors, the Second Rome of the popes, and the Third Rome of the bureaucrats, which began abruptly when the city was thrust into the modern era after being made capital of the recently united Italy. It was preceded in that role by Turin, which had been capital from 1861 until 1865, then by Florence, whose tenure lasted six years. But symbolically and geographically, Rome seemed the natural choice for Italy's permanent capital.

There was just one thing standing in the way: Rome was still nominally in the hands of the papacy. Until, that is, the city was finally taken by force when the Italian army breached Porta Pia on September 20, 1870. Pope Pius IX retreated into the papal compound, famously declaring himself "prisoner in the Vatican": a self-imposed exile and title that successive popes maintained for sixty years, in futile protest over the loss of their historical seat. Rome's destiny

Fig. 124

The Monument to King
Vittorio Emanuele II.
Photo: Paolo Costa Baldi
/ Wikimedia Commons.
CC BY-SA 3.0: https://
creativecommons.org/
licenses/by-sa/3.0/legalcode.

was sealed soon after, with the signing of official legislation designating its new status as Italy's capital on July 1, 1871.

There ensued a series of herculean and often contentious efforts to retrofit modern planning onto the 2,700-year-old city. In the immediate aftermath of 1870–71 and for decades afterward, various proposals were put forth for how to update Rome and equip it for its new role. The city's still largely medieval infrastructure required dramatic transformation to carve out the modern roads and spaces deemed fitting for a great capital, to make room for new ministries and transport systems, not to mention the growing population, which surged tenfold in the period discussed in this chapter, from just over 200,000 in 1870 to more than two million by 1960.

However you frame it, Rome experienced some of its most pronounced growing pains after 1870. Few visitors today grasp the extent to which the city they see is a product of the modern era—from the prioritization of certain ancient ruins and sites over others to the presence and spotlighting of select modern monuments, the framing of urban vistas, and the balance of residential, governmental, and archaeological zones, as well as the pattern and flow of city streets.

The maps in this chapter have been chosen precisely because they provide windows onto the critical process of change that accelerated after Rome became capital, paving the way toward its current state. Rome's official master plan of 1883 allows us to trace the early stages of conceptualization and implementation, while later maps disclose the extent to which the promise of that plan materialized (or failed to). Along the way, we will also see the evolution of Rome's mass transit system and the major strides in urban planning that accompanied the 1960 Olympics, an ephemeral event but one that provided yet another occasion for lasting urban transformation.

From the beginning, the evolution of *Roma capitale* proceeded at a dizzying pace. In the last chapter we saw new avenues and entire neighborhoods begin to appear in tourist maps from the 1870s and 1880s. These years also saw the birth of Rome's most conspicuous modern landmark: the monument everyone loves to hate, dedicated to the first king of united Italy, Vittorio Emanuele II (whose name is sometimes anglicized to Victor Emmanuel II). Today, the massive structure, which also houses Italy's tomb of the unknown soldier, dominates the traffic vortex that is Piazza Venezia while providing a prominent visual terminus to Via del Corso (fig. 124). Officially known as the "Altare della Patria," or altar of the homeland, and familiarly as the Vittoriano, it was conceived as a secular shrine to the Italian state—and an attempt to outshine the Christian shrine of St. Peter together with the papal authority it embodied.

The Vittoriano was begun in 1885 and inaugurated in 1911, when a celebratory dinner for twenty-four dignitaries was held in the belly of the huge bronze equestrian statue of the king—a statue emulating that of Marcus Aurelius on the adjoining Campidoglio but many times larger—that formed the centerpiece of the monument. The structure was not fully complete, however, until 1930, when the crowning quadrigas (bronze four-horse chariots, driven by winged victories and modeled after ancient prototypes) were hoisted into place at either end.

The monument has been mocked as the "typewriter" for its resemblance to a vintage Olivetti, or alternatively as the "wedding cake" for its tiered structure clad in a profusion of frothy white marble, "imported" from distant Brescia— where, it so happens, the minister of public works at the time of its construction had close ties. Today, this gargantuan neoclassical pastiche can be appreciated as an unintentionally ironic, almost postmodern rehash of ancient Roman motifs, or in all seriousness for its military museum and temporary exhibitions, as

Fig. 125

Vintage photograph of the Tiber embankments. Oregon State University Special Collections & Archives / Wikimedia Commons.

Rome Enters the Modern Age

well as for its stunning views high above the city (the elevator trip to the top level is worth the extra euros). From there, it is easy to forget the structures and topography that were flattened to make way for the pompous colossus underfoot.

But the project that most profoundly altered the city's form for the sake of modernization—and that best encapsulates the fraught but intimate relationship between urban improvement and loss—was less flashy: the construction of the Tiber embankments, which began in 1875 and continued into the early twentieth century (fig. 125; see also fig. 126). Planning began after the disastrous flood of 1870, just one in a very long line of floods that had wreaked havoc on the city. Hoping to rid Rome of these periodic cataclysms once and for all, city officials decided on a drastic course that ultimately involved shoring up the riverbanks with massive stone retaining walls that regulated the Tiber's course and held it in check. The project necessitated the redefinition of adjoining streets and the demolition of a multitude of historical constructions that lined the river.

Significant portions of the old Jewish Ghetto were sacrificed, as was the upriver port known as Ripetta, an elegant, swooping urban set piece from the early eighteenth century designed by Alessandro Specchi. A series of wide tree-lined thoroughfares stretching for miles was laid out along the embankments, flanking both sides of the river. Known collectively as the *lungotevere* (literally, along the Tiber), these high-volume roadways cut through the center of Rome. If the city has since been spared the devastating floods of the past, the Tiber has been lost as a living presence in the city. Sunken and diminished, it is no longer Rome's pulsing lifeblood. That role has been taken over by cars.

When Rome became capital, various scenarios were entertained for how best to interweave the bureaucratic infrastructure into the urban fabric. One plan called for concentrating the political apparatus on the Quirinal Hill, along Via XX Settembre—the old Via Pia, renamed to honor the day Rome was taken for Italy. At the avenue's western end, the Palazzo del Quirinale, an alternate papal palace from the late sixteenth century, was seized and repurposed to become the royal residence—after World War II, the residence of Italy's president. The adjoining eighteenth-century Palazzo della Consulta, another requisitioned papal holding, became the Ministry of Foreign Affairs in 1874 (and later the Constitutional Court). Over the next several years, the Ministries of War and of Finance established headquarters further up the street in massive purpose-built structures.

Additional government buildings were constructed along Via Nazionale, parallel to Via XX Settembre and one block over. A grand space for official exhibitions, the Palazzo delle Esposizioni designed by Pio Piacentini, was inaugurated in 1883. A bit further down the street, work began on the enormous headquarters for the Bank of Italy, designed by Gaetano Koch, in 1885.

The notion of a dedicated governmental quarter did not proceed much further. Instead, ministries and other official buildings came to be scattered across the city in piecemeal fashion, comprising a mixture of structures old and new. Palazzo Montecitorio, a lavish palace north of the Pantheon designed by Bernini in the 1600s, became the seat of the Chamber of Deputies. The Italian

Fig. 126.

Aerial view of the Palazzo di Giustizia (Palace of Justice), Rome. Photo: Alinari / Art Resource, NY.

Senate was installed nearby, in Palazzo Madama, a Renaissance palace built for the Medici family. The behemoth Palazzo di Giustizia, designed by Gugliemo Calderini to house Italy's supreme court, was begun in 1889 on a site east of the Vatican, across the Tiber from the center (fig. 126).

National capitals also entail foreign embassies and consulates. In Rome, that number is greater than in many other capitals, because Vatican City—which became the world's smallest independent nation with the signing of the Lateran Treaty in 1929—does not actually have room for embassies within its precincts, so diplomatic missions to the Holy See are located in Rome. Currently, the city is home to almost ninety embassies to the Vatican, as well as more than 140 to Italy, not to mention miscellaneous others (embassies to the Sovereign Order of the Knights of Malta, for example, are also located in Rome). Many of these buildings occupy extremely valuable real estate. The great Renaissance Palazzo Farnese became home to the French embassy in 1936, while the Brazilian Embassy has occupied the seventeenth-century Palazzo Pamphilj, on Piazza Navona, since 1920.

By the early twentieth century, Rome was buzzing with activity and had been for decades. Although the city had to back out of hosting the Olympics of 1908 for reasons that remain somewhat obscure, it made up for that disappointment three years later with an international exhibition intended to show off the revamped capital and celebrate the fiftieth anniversary of statehood. The mood, however, was not to last. Italy's finances were already heavily strained at the onset of World War I, and the situation degenerated after Italy entered the conflict in spring 1915—a decision that many citizens did not support. Nominally among the victors when the war ended in 1918, Italy emerged in shambles.

Rome Enters the Modern Age

Fig. 127

Mussolini on the balcony
of Palazzo Venezia, 1932.
Photo: Adoc-photos / Art
Resource, NY.

Inflation was steep and the cost of living had doubled, labor issues were wide-spread, significant discontent prevailed in various sectors of the population, and true unity still seemed illusory. Conditions were ripe for nationalism to rear its ugly head.

The Fascists swept into power behind the charismatic dictator-in-the-making Benito Mussolini in 1922. Recognizing Rome's potent symbolism and seeking to align it with Fascist ideology, Mussolini and his ministers launched another key chapter in the city's remaking. An ideal (or cult) of *romanità*—Romanness—was key to the regime's image. Mussolini styled himself a new emperor and was determined that his city would match, subsume, and ultimately outdo that of antiquity. "In five years," he wrote, "Rome must appear marvellous to all the peoples of the world; vast, orderly, powerful, as it was in the time of the first empire of Augustus."

While in many respects Mussolini's interventions continued policies and ideas already in place since 1870, he pushed the destructive impulses to an extreme level, often personally, proudly wielding the pickax that struck the first blow to existing structures. Fascist urbanism involved showcasing certain prominent ancient monuments by isolating them—"cleansing" them of later, usually medieval, accretions; modernizing Rome's infrastructure, particularly its streets; and creating large open spaces ideal for the performance of militaristic ceremonials.

Much of this activity and spectacle revolved around Mussolini's headquarters at Palazzo Venezia, from the balcony of which he famously harangued the crowds in the piazza below (fig. 127). Already partially cleared for the construction of the Vittoriano, this space was now swept fully clean to showcase Fascist Rome and its leader, "Il Duce." At Mussolini's behest, a thriving working-class neighborhood between the Vittoriano and the Colosseum was razed to make room for the wide, straight avenue—ideal for staging military parades—that came to be known as Via dell'Impero (Street of the Empire, now Via dei Fori Imperiali). Former residents were relocated to new, mostly squalid municipal settlements on Rome's outskirts.

On either side of Via dell'Impero, the Roman and imperial fora were cleared of medieval and Renaissance residue to place them in a pristine state of majestic isolation, as was the adjacent Capitoline Hill. The close proximity of Mussolini's base at Palazzo Venezia to these newly sterilized archaeological sites was far from accidental: rather, it was all part of his master plan to link Rome's glorious past to its present and future under Fascism.

Equally dramatic changes occurred elsewhere. The Mausoleum of Augustus, near the Tiber along Via di Ripetta, was stripped of symbiotic postantique structures, then surrounded by a rationalist nest of Fascist architecture, as well as the transplanted Augustan Ara Pacis, or Altar of Peace, enclosed in its own modernist box. In the Vatican Borgo, meanwhile, a whole city block was demolished to create a monumental approach to St. Peter's, Via della Conciliazione (fig. 128). This Fascist initiative was meant, in part, to project the recent rapprochement between the regime and the papacy. On a functional and aesthetic level, the new street did provide better access for pilgrims while opening a scenographic vista from the river to the basilica. At the same time, a whole residential quarter, including many architectural monuments, fell to the wrecking ball.

In addition to clearing away later accretions to showcase ancient landmarks, Mussolini also spearheaded the creation of new monumental centers to promote various facets of life under his regime: physical health, education,

Fig. 128

View from the roof of St. Peter's Basilica of St. Peter's Square and the Via della Conciliazione. Photo: David Iliff / Wikimedia Commons. CC BY-SA 3.0: https:// creativecommons.org/ licenses/by-sa/3.0/legalcode.

Fig. 129 Comune di Roma, *Piano regolatore e di ampliamento della città di Roma*, 1882. By permission of the Archivio Storico Capitolino–Sovrintendenza ai Beni Culturali di Roma Capitale, Cart. XIII, 109.

S.P.Q.R.

PIANO REGOLATORE

E DI

AMPLIAMENTO DELLA CITTÀ

DI

ROMA

*approvato dal Consiglio comunale nella
seduta del 26 Giugno 1883 in esecuzione della
legge sul concorso dello Stato nelle opere
edilizie nella Capitale del Regno*

Leggenda

and entertainment. The suggestively named Foro Mussolini (Forum of Mussolini), a sporting venue meant as a place to strengthen the bodies of Fascist youth, opened in 1933. Rome's Città universitaria, or university city, aimed to strengthen their minds. This large complex on the Esquiline enrolled its first students in 1935. Cinecittà, the cinema city, opened its doors as a Fascist propaganda machine in 1937.

World War II put a stop to all this activity, but many of Mussolini's projects were revived in the postwar years, living on under different auspices and circumstances. Città universitaria is still the flagship campus of Rome's university, "La Sapienza," which enrolls over 140,000 students. Cinecittà might have had its heyday in the 1950s and 1960s, but it remains operational today. Movies and television series produced there in recent decades include Martin Scorsese's *Gangs of New York* (2002) and HBO's *Rome* (2005–2007) and *The Young Pope* (2016). Similarly, the Foro Mussolini, rechristened the Foro Italico, still hosts sporting events as well as concerts.

Overall, Mussolini and his minions eagerly stripped away much of Rome's historic and living fabric in the name of convenience and progress—not unlike a different kind of dictator, Robert Moses, did in New York City from the mid-1930s to the 1960s (his cult: the automobile). Mussolini was one in a long line of authoritarian powers who made their mark on Rome, but at its core, the city has also shown time and again that it has a mind of its own. Its urban form reflects thousands of years of push and pull between the will of powerful individuals to impose their vision, on the one hand, and the collective will of inhabitants to shape a city organically through time and use, on the other.

Italy and its citizens emerged utterly devastated from World War II and Mussolini's disastrous alliance with Hitler, but Rome's physical fabric was largely intact. Slowly, over the following decade and a half, the nation would rebuild. In 1946 Italy became a republic, and in 1949 it joined NATO. By 1950, with the aid of the Marshall Plan, the country had stabilized financially, and by 1960, it was in the midst of an "economic miracle." Rome, which became the birthplace of the European Union in 1957 with the signing of two treaties, was ready for its close-up on the world stage. This chapter traces the city's fraught path to modernity in maps from 1870 through the Olympics of 1960—a major, televised event meant to showcase the triumphant return of Rome, finally standing for Italy as a whole.

2,500 Years in, a Master Plan for Rome

City officials wasted little time after Rome was claimed for Italy in 1870 to draw up a *piano regolatore*, or master plan for the city's future. As a result, Rome was accorded a treatment it had never previously been granted in its millennia of history: a blueprint for development. Sixtus V had come closest in the sixteenth century, but his scheme was focused on one district, the underdeveloped greenbelt. Now, city officials were charged with creating a comprehensive design that would address the whole spectrum of urban challenges—circulation, housing,

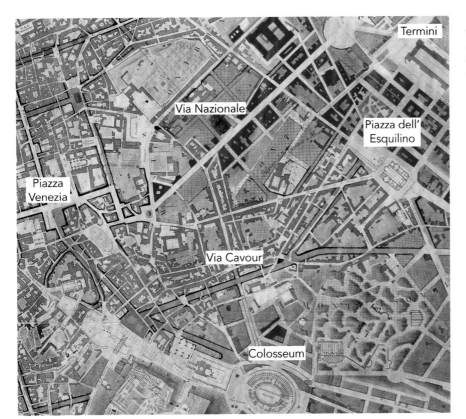

Fig. 130

Piano regolatore, 1883,
detail showing the Quirinal
and Esquiline

and public life—encompassing the renewal of Rome's historical core as well as the city's expansion into new quarters, considering the needs of the city's permanent and transient populations for the foreseeable future.

The first official plan emerged in 1873. The brainchild of urbanist Alessandro Viviani, it was never fully ratified or funded, much less implemented. As the city's population surged in that decade, private and unregulated building activity rushed in to fill the gap. By 1880, it became clear that Rome's growth could not be allowed to continue unchecked. A commission was created to vet a new version of the 1873 plan in hopes of getting a handle on Rome's haphazard growth. Viviani's updated design, illustrated here (fig. 129) in a presentation version dated 1882, was ratified in early 1883. In its straightforward color coding we see evidence of so many drastic changes that were planned or underway.

A category called "public boulevards" consisting of new or expanded streets is depicted in pale orange. Among the most important is Via Nazionale at right, the eastern branch of a new east-west axis through the city, which created a perpendicular to the north-south Via del Corso while linking Termini train station to the Vatican (fig. 130). It was begun before unification, in the 1860s. Originally, it stretched only a few blocks on the Quirinal, from the piazza fronting the Baths of Diocletian to Via delle Quattro Fontane (part of Sixtus V's network of streets). After 1870, it was extended to run further into the center, toward the Markets of Trajan and imperial fora, making a sharp right as it moved downhill to avoid running directly into the Column of Trajan, then a sharp left again to

Rome Enters the Modern Age

Fig. 131

Piano regolatore, 1883,
detail showing Corso Vittorio
Emanuele II and monu-
ments lining its course

skirt the north side of Piazza Venezia. There, it flowed into Via del Plebiscito,
which in turn fed into the new Corso Vittorio Emanuele II.

Darker orange is used to signal provisional streets and spaces that required
demolition of existing urban fabric. As indicated by the orange patches branch-
ing out from Piazza Venezia to the southeast, Viviani's plan called for the raz-
ing of part of that zone to create a more open space and a wider street leading
toward it from the Colosseum. The intervention as depicted here is less radical
than what would come to pass once the Vittoriano was in progress and of course
later, under Mussolini.

Another avenue that would become a reality is evident in the uninterrupted
line of orange originating at Piazza dell'Esquilino on the north side of Santa
Maria Maggiore, angling to the west as it moves downhill to intersect with the
Forum. This street was an extension into the center of Via Cavour, laid out in
the 1880s to connect Termini train station and Piazza dell'Esquilino.

Corso Vittorio, Viviani's most successful intervention, is shown on the map
zigzagging its way through the heart of the old city (fig. 131). The numerous
orange demolitions along its route could be likened to obstructions swept out
of the way by a flash flood through a gorge. But in truth this thoroughfare was
remarkable for its delicate balance of preservation and modernization. While
some buildings deemed to be of lesser importance fell victim to its route, it managed to shimmy along an obstacle course that dodged a number of Renaissance
and Baroque monuments—the church of the Gesù, the Palazzo Massimo alle
Colonne, the Palazzo della Cancelleria, Chiesa Nuova, and others—then on in
the direction of the Tiber and Vatican via the Ponte Sant'Angelo (the bridge

Chapter Nine

leading to the Castel Sant'Angelo) and the planned Ponte Vittorio Emanuele II, completed in 1910.

Many spots along the Tiber, where the embankments were in progress, are also yellow. The river is additionally flanked all along its course by evenly spaced, uninterrupted double lines of pink, a color used to signal planned constructions—in this case, the paths of the *lungotevere* boulevards. The master plan also called for eight new bridges over the Tiber; this many and more would be realized before 1945.

In the map, dark pink is used to signal projected quarters. Most are concentrated on the south side of the city: on the Aventine, near Porta San Paolo, and in Testaccio. The largest is Prati, north of the Vatican at the map's upper left, which is portrayed as a gridded blank slate (fig. 132). Its anchor, the Palazzo di Giustizia, is depicted but not yet begun. One of the few planned neighborhoods that did not encroach on existing settlement to any appreciable extent, Prati was also one of the few to lie outside the Aurelian Wall. Otherwise, the plan of 1883 did not consider the scenario of the city outgrowing its ancient perimeter. In retrospect this seems shortsighted, but in the 1880s there was still a lot of open space, and a lot of building activity, within those walls.

A dark reddish orange color, meanwhile, is used for new quarters whose construction was already underway. The area of the Esquiline and around Termini, at right, is blanketed in this hue—which outlines large city blocks that contrast with the city center as much for their geometric homogeneity as for their color scheme (fig. 133).

Fig. 132

Piano regolatore, 1883, detail showing Prati di Castello

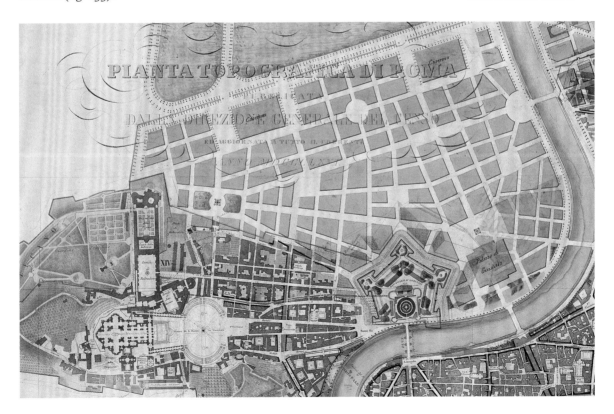

Rome Enters the Modern Age

Fig. 133

Piano regolatore, 1883,
detail showing new quarters
on the east side of the city

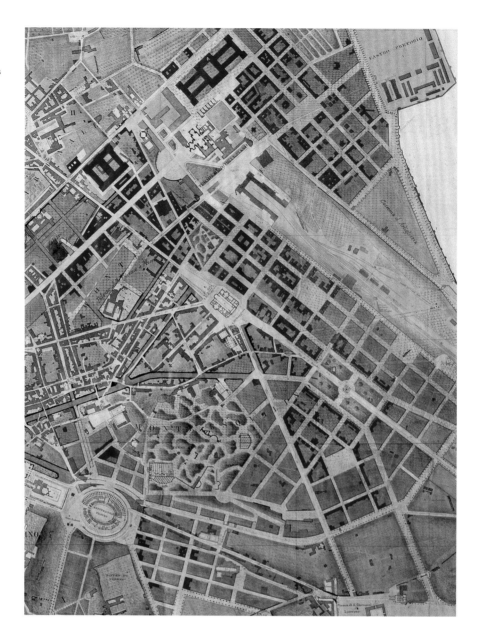

The master plan of 1883 was obviously prescriptive, and much of it was implemented over the next two decades. It was not a solution for Rome's future, however, nor did it allay the corruption and property speculation that ran rampant, thwarting the realization of any coherent vision. Municipal and governmental officials were often at cross-purposes. For better or worse, for Rome there would be no Baron Haussmann—the mastermind responsible for the wholescale revamping of Paris, a process just concluded in 1870. Rather, much of Rome's development consisted of halting halfway measures interspersed with grandiose interventions that were far from sensitive to the existing fabric, the realization of which involved demolishing precious relics of the past or bulldozing functional, longstanding neighborhoods.

Little attention, meanwhile, went to resettling the scores of residents displaced to make room for such undertakings, so housing became a perpetual problem. That said, along with the losses there were gains. Certain steps taken to renew Rome's infrastructure did improve conditions in the city, such as circulation and sanitation. More than at any earlier time, demolition and new construction went hand in hand.

When Trams Ruled Rome

As Rome's population surged after 1870, the public transport system grew apace, with no real plan in place and no central authority to guide it. Beginning in 1845, passengers had moved through the city in horse-buses, essentially large coaches. Run by a private company, the Società Romana Omnibus, these relatively informal conveyances did not operate according to a consistent timetable. In 1877, they began to be replaced by horse-drawn trams. These streetcars were larger, and they ran along steel rails that reduced friction, facilitating a smoother ride and allowing the horses to pull greater loads. In 1880, a line running between Termini and Piazza Venezia was inaugurated. Operated by the same company, now rechristened the Società Romana Tramways e Omnibus (SRTO), it ran with regular service along Via Nazionale. After 1890, the trams gradually shifted to electric power. It was an exceptionally popular system, and Rome came to be crisscrossed by a tangled web of tracks run by competing, although often complementary, companies.

A municipal tramway came into being only in 1909, initially existing alongside the private lines, then gradually subsuming them after World War I. In 1927, the system was placed under the supervision of a new agency, the Azienda delle Tramvie e Autobus del Governatorato di Roma (ATAG). It was a sprawling and chaotic network, consisting of hundreds of vehicles and fifty-nine tram lines running over approximately 250 miles of track. ATAG planners were already working toward consolidating the system when a radical reorganization was ordered in 1930. The resulting reform called for the elimination of overlapping lines and the drastic reduction of track. Most important, it banned trams entirely within the Aurelian Wall, replacing them with a system of buses. A circular tramline was put in place to run along the perimeter of the city center, linking the intra-urban bus system with the extra-urban tramways.

The reform was part of Mussolini's overarching program to shape Rome in his vision of a twentieth-century metropolis. Ostensibly, the motives for banning trams were to reduce congestion within the city, facilitate maintenance by employing a more reliable bus system, and make passage easier for motor vehicles, which numbered about 30,000 (and counting) in 1930. More than that, Mussolini considered trams old-fashioned, preferring a more modern autovehicular transport system for his capital.

The highly detailed map illustrated here (fig. 134) captures Rome in 1925, at the height of the tram era. It was included in a German version of a guidebook

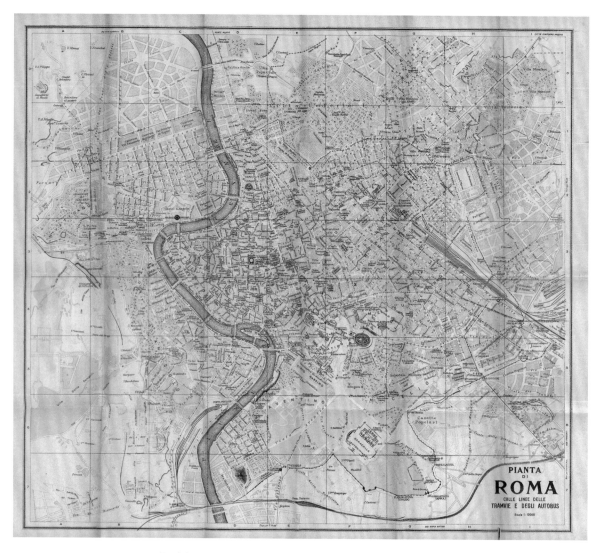

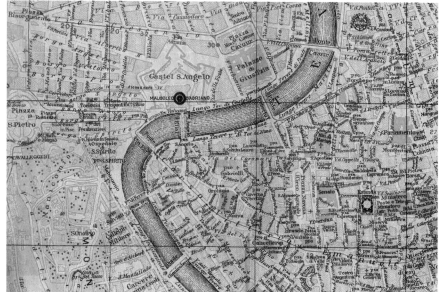

Fig. 134

Above: *Pianta di Roma con
le linee delle tramvie e degli
autobus*, Rome: Enrico
Verdesi, 1925. By permission
of the Archivio Storico
Capitolino–Sovrintendenza
ai Beni Culturali di Roma
Capitale, 19056.

Fig. 135

Below: *Pianta di Roma con
le linee delle tramvie*, detail

by Aristide Tani, issued by the Roman publisher Enrico Verdesi. Tram lines, in red, course through the city, creating a delicate circuitry throughout that only begins to hint at their pervasive presence (fig. 135). Darting everywhere, the network circles the Pantheon, skirts the curved ends of Bernini's colonnades for St. Peter's square, converges along Corso Vittorio, crosses the new Ponte Umberto I toward the Palazzo di Giustizia, then creeps around that bulky mass to circle the new Piazza Cavour and head into Prati, now a thriving quarter— and so on and so forth.

Tramlines run up and down new streets, whose development went hand in hand with their own. Via Cavour is now a reality, providing a direct link between Termini and Piazza Venezia, although Via dell'Impero has yet to appear. Similarly, demolitions had not yet begun for Via della Conciliazione leading up to St. Peter's, so in that zone the trams march up one street and back down another, their parallel paths flanking the row of buildings that would be razed a few years later to make room for the monumental approach to the basilica.

Each thin red strand of tramline on the map corresponds to a whole sensory experience of Rome in the 1920s that is hard to recapture: thronged, clogged streets, bells clanging, crowds jostling to climb on board. The bells would be replaced within decades by the horns and screeching brakes of the automobiles that were only beginning to make their presence felt in the city.

An Olympic City, and a New Beginning

When Rome won its bid to host the 1960 Olympic games in 1955, it was seen as an opportunity for Italy to reintroduce itself to the world after decades of turmoil. Not coincidentally, these games were to be the first televised internationally. Italy had made a remarkable recovery from its nadir during Fascism and World War II, and Rome, for its part, had been modernized in countless ways over the preceding century. The street network was vastly different than it had been in 1870. New modes of transport had been introduced into the city. Termini had been remade, the new station inaugurated in 1950. Many of Rome's ancient and Renaissance landmarks had been curated into an open-air museum. In sum, the city was equipped with the infrastructure and monumentality of a capital.

Rome also had many of the trappings of an Olympic host city already in place or in progress—ironically, thanks to Mussolini. In contrast to so many later host cities, which have taken on billions of dollars in debt to create glittering new venues only to see them fall into disuse, Rome managed to repurpose or expand existing structures, which remain active centers today. At the northern end of the city, the Foro Mussolini—Il Duce's grandiose sporting complex and monument to Fascist fitness, designed by Enrico del Debbio—was renamed the Foro Italico and adapted to become one of two main centers of activity. In preparation for the Olympics, its Fascist program of slogans and statues was tempered, some existing structures were updated, and new ones were added.

Fig. 136

Opposite: *Roma Olimpica*
MCMLX, Rome: Enrico
Verdesi, 1960. Private
collection.

The other principal venue was south of the city, where Mussolini had broken ground on a new monumental quarter in 1935. Named Esposizione universale di Roma '42 (or E42), later EUR, it was intended to be the site for a world's fair in 1942 and then a permanent feature of Rome's extended urban matrix. Little progress was made on this planned economic and cultural center before World War II put an end to the exhibition before it started, along with so many of Mussolini's other ambitious projects. Twenty years later, construction was eagerly restarted for the 1960 Olympics.

Long before sustainability was a buzzword, Rome's Olympic planners were committed to recycling existing buildings or to implementing new infrastructure that would have enduring relevance. Housing for athletes in the Olympic Village constructed on Via Flaminia just south of the Foro Italico was meant to become a permanent residential quarter for civil servants following the games. Designed by a team including Adalberto Libera and Luigi Moretti, this neighborhood was a planned community including not only apartment buildings but also a church, shops, restaurants, bars (of the *caffè*-serving variety), and schools.

The Olympics were also a catalyst to continue improving public transport. Rome's main airport, Leonardo da Vinci, which lies southwest of the city at Fiumicino, was built in anticipation of an influx for the event, but it was also correctly envisioned to have more lasting purpose. A new subway, known as Metro B, opened in 1955 to connect Termini with EUR. It is another example of a project begun under Fascism and resurrected in preparation for 1960, for it was originally conceived in 1937. Major roadways were also completed for the games, most notably the so-called Via Olimpica that linked Foro Italico to EUR, in the process tragically slicing through the Villa Doria Pamphilj—approximately 450 acres that still belonged to one of Rome's most prominent old families. Acquired by the city in the following decade, that land was converted into Rome's second-largest (albeit strangely bisected) public park.

Just as that artery took a tangential route west of Rome, so too planners wisely chose to situate the main sporting venues north and south of the city, thereby avoiding the historic center—with a couple of noteworthy exceptions. Within the walls, athletic competitions were staged at the Baths of Caracalla and the Basilica of Maxentius. The marathon, for its part, began at the base of the Capitoline, with runners eventually circling back into the center to cross the finish line at the Arch of Constantine next to the Colosseum. These choices brilliantly called attention to Rome's ancient patrimony—to its three millennia of history and culture—while keeping observers from dwelling for long on the city's uglier recent chapters.

This official map for the 1960 games (fig. 136) spotlights the Olympic sites in bright red over a basic street map of the city, extended southward to reach EUR. Visible toward lower left, this formal Fascist utopia is today a thriving residential, industrial, and cultural center, housing the Italian national archives and the Museo della Civiltà Romana (Museum of Roman Civilization)—also known as the home of Gismondi's *Plastico* (fig. 29). EUR also houses some of Fascist Rome's most evocative architectural monuments, including the Palazzo dei Congressi—a conference center—with its distilled references to the

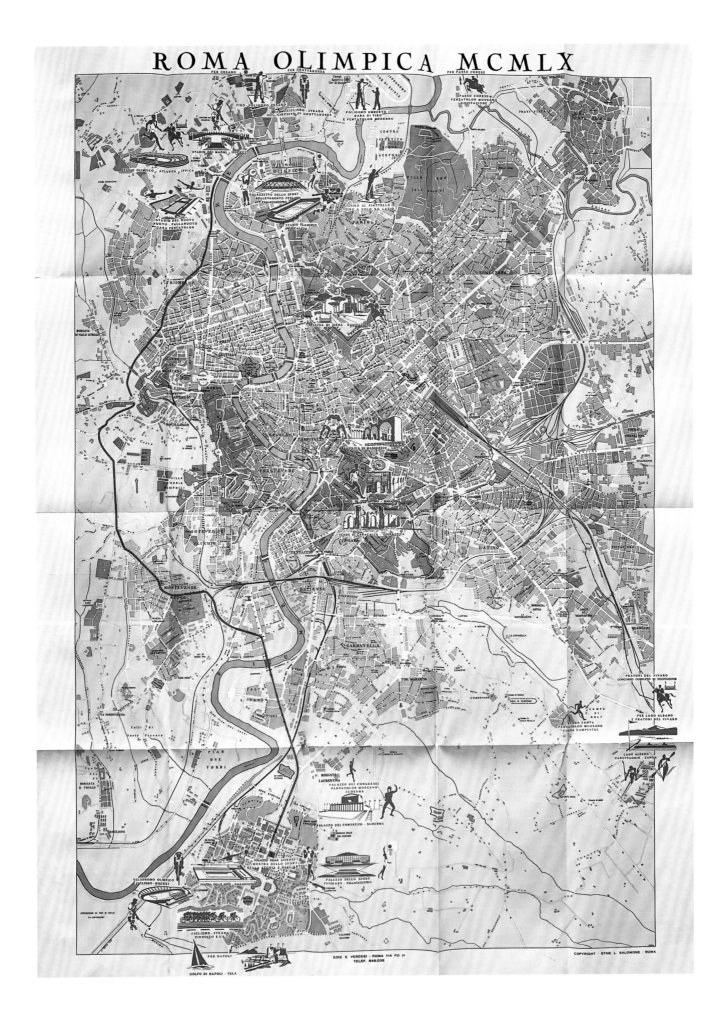

ROMA OLIMPICA MCMLX

Pantheon and other symbols of Roman greatness, and the striking Palazzo
della Civiltà Italiana, known familiarly as the Colosseo Quadrato (or Square
Colosseum) for its monolithic rectangular form perforated by deep arches cast-
ing dramatic shadows (fig. 137). Olympic structures added to the original plan,
which was largely the work of Mussolini's favorite architect Marcello Piacentini,
include Pier Luigi Nervi's influential Palazzo dello Sport, a velodrome, and a
pool for water polo.

On the map, EUR is linked to the Foro Italico on the northern end of the city
by a thick red line, Via Olimpica, curving around and up the left side. Venues
depicted at its upper reaches include the Stadio Olimpico, which was adapted
to host track and field as well as the opening and closing ceremonies. Below it
is a new swimming complex and to the right the Stadio dei Marmi, part of the
original Foro Mussolini, used in 1960 for hockey. Across the river, on the city
side, is the Olympic Village as well as the smaller of Nervi's two arenas, the
Palazzetto dello Sport, used for basketball. Close by is the Stadio Flaminio, a
soccer field built in the 1910s.

Like a secular Jubilee, the Olympics of 1960 functioned as a temporary gathering that spurred improvements benefitting legions of short-term visitors as

well as Rome's permanent population. The event lasted all of two weeks, but it was a resounding success for the city, projecting the image of a modern, fashionable, vibrant capital to a worldwide audience. Well planned and executed, the games were instrumental in helping Rome—and the nation as a whole—to rehabilitate its damaged image and to tie a neat bow on the first, tumultuous century of Italian unity.

FURTHER READING

Agnew, John. "The Impossible Capital: Monumental Rome under Liberal and Fascist Regimes, 1870–1943." *Geografiska Annaler. Series B, Human Geography* 80, no. 4 (1998): 229–40.

Atkinson, David, and Denis Cosgrove. "Urban Rhetoric and Embodied Identities: City, Nation, and Empire at the Vittorio Emanuele II Monument in Rome, 1870–1945." *Annals of the Association of American Geographers* 88, no. 1 (1998): 28–49.

Brennan, T. Corey. "The 1960 Rome Olympics: Spaces and Spectacle." In *Rethinking Matters Olympic: Investigations into the Socio-Cultural Study of the Modern Olympic Movement*, ed. R. K. Barney, J. Forsyth, and M. K. Heine, 17–29. London, ON: International Centre for Olympic Studies, 2010.

Costa, Frank. "Urban Planning in Rome from 1870 to the First World War." *GeoJournal* 24, no. 3 (1991): 269–76.

Ghirardo, Diane. "From Reality to Myth, Italian Fascist Architecture in Rome." *Modulus* 21 (1991): 10–33.

Kallis, Aristotle. *The Third Rome, 1922–43: The Making of the Fascist Capital.* New York: Palgrave Macmillan, 2014.

Kirk, Terry. *The Architecture of Modern Italy.* 2 vols. New York: Princeton Architectural Press, 2005.

Kirk, Terry. "Framing St. Peter's: Urban Planning in Fascist Rome." *Art Bulletin* 88, no. 4 (2006): 756–76.

Kostof, Spiro. "The Drafting of a Master Plan for 'Roma Capitale': An Exordium." *Journal of the Society of Architectural Historians* 35, no. 1 (1976): 4–20.

Kostof, Spiro. *The Third Rome, 1870–1950: Traffic and Glory.* Berkeley, CA: University Art Museum, 1973.

Maraniss, David. *Rome 1960: The Olympics that Changed the World.* New York: Simon & Schuster, 2008.

Painter, Borden W., Jr. *Mussolini's Rome: Rebuilding the Eternal City.* New York: Palgrave Macmillan, 2005.

Tocci, Walter, Italo Insolera, and Domitilla Morandi. *Avanti c'è posto: Storie e progetti del trasporto pubblico a Roma.* Rome: Donzelli, 2008.

Chapter Ten

Rome Past, Present, and Future

"Rome of the future will be a metropolis of contiguous centralities, tied together by the protected environmental systems and the planned mobility network."

Beatrice Bruscoli, architect

"With an inefficient service, a chaotic governance model, and continuous scandals, Rome's unresolved transportation crisis forged a city of social fragmentation, environmental degrade, and economic stagnation. Today, Rome lags far behind its European counterparts in the implementation of a sustainable urban agenda, where mobility policies play a fundamental role."

Cosima Malandrino, urbanist

Inside the Aurelian Wall, Rome has undergone little physical change since 1960. Outside that boundary, however, it is unrecognizable: a sprawling metropolis, with sprawl being the operative word. As of 2019, greater metropolitan Rome has a population of more than four million. The center is and will always be Rome's tourist magnet and political hub, but actual Romans are increasingly priced out of it, especially with services like Airbnb appropriating more and more properties for the lucrative and controversial business of short stays.

Tourism in general is as much of a threat as a boon to the city these days. The number of visitors has skyrocketed in the past few decades, bringing revenue to the city's coffers but at times stretching the limits of Rome's capacity to host them. Mass transit, meanwhile, is not what it should be, while car use is correspondingly out of control. The city is plagued with some of the worst traffic problems and pollution levels in Europe.

In a sense, Rome is a victim of its own success. Can or should it continue to grow in such fashion, and how can the city move sustainably into the future? This chapter brings the book to a close—and Rome up to the present—with several maps relating to projects that claim to promote a critical path forward. The master plan of 2008 and the official public transport planning map are intended to work in tandem to safeguard Rome's ongoing, healthy development, but the motives behind them have been questioned, and there is good reason to doubt their feasibility if not their good faith.

It is certainly possible to point to positive developments since the turn of the millennium. There have been admirable efforts to reintegrate the Tiber into the city through artistic and cultural initiatives. And for the first time since antiquity—unless you count Fascist-era structures—Rome has become a home to cutting-edge architecture. Richard Meier's Ara Pacis Museum, which opened in 2006 after much hype and criticism, is a rare example in the old city center. The same architect is responsible for the Jubilee Church (2003) in the Tor Tre Teste district east of Rome. The Flaminio area north of Piazza del Popolo, near the Olympic Village, houses Renzo Piano's Auditorium Parco della Musica (2002) as well as Zaha Hadid's contemporary art museum, MAXXI (Museo nazionale delle arti del XXI secolo, 2010). In the Fascist-planned district of EUR south of the city, the budget-busting convention center dubbed "The Cloud" by Studio Fuksas—an architectural wonder or folly, depending on whom you ask—was finally completed in 2016 after years of delays.

At the same time, even if physical changes to the urban core are few, the character of the city has changed in the last half century and more. Rome has become more global. Immigrants—although not always welcomed with open arms—hail from Bangladesh, China, the Philippines, Eritrea, Sudan, Senegal, and elsewhere. The area around Piazza Vittorio Emanuele II is a bustling multiethnic neighborhood. Restaurants offer cuisines from all around the globe, so it is easy to branch out from pasta if you wish (although why wouldn't you want a delicious carbonara?).

Green space in Rome is a mixed bag. In the unregulated building boom of the late nineteenth century, many gardens that had belonged to private villas were transformed into apartment blocks. But those that were preserved and made public—the Villa Ada, Villa Borghese, and Villa Doria Pamphilj west of the center—are delightful, expansive urban oases. That said, maintenance of the city's parks is spotty. The Villa Borghese, although recent exposés have lamented its condition, is better kept than, say, the Parco del Colle Oppio above the Colosseum. Recent reviews on tripadvisor.com are headlined "VERY Sketchy Park" and "be careful." On a nice spring evening, however, there as elsewhere you will see parents pushing strollers, kids playing soccer, and other benign scenes of park and piazza life in Rome.

Still, to many observers, the city is dirtier and grittier than ever, even teetering on the brink of collapse. As of this writing, trash piles collect on Rome's

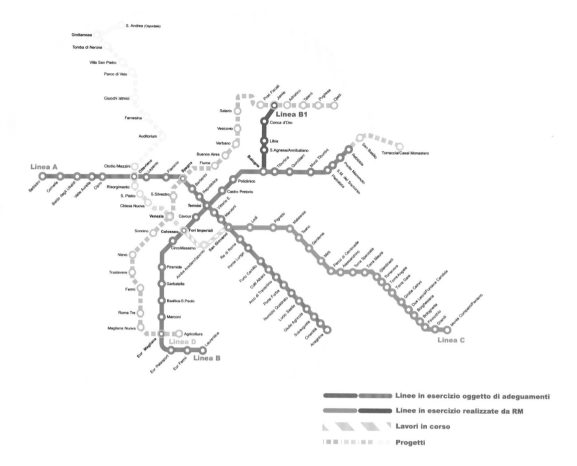

Fig. 138

Metropolitane di Roma, provisional map of Rome's metro system, 2018. © Roma metropolitane.

most storied streets and squares. Like the Parco del Colle Oppio, many of the smaller city parks are garbage-strewn weed pastures with overgrown, graffitied, sometimes dangerous playgrounds where shards of glass litter the ground and dogs do their business as their owners look the other way. Rome's transport system is a mess, and city buses infamously burst into flames on a regular basis.

The current mayoral regime tends to be blamed for this latest ugly chapter, with a recent *New York Times* article proclaiming "Rome in Ruins" and calling the city a literal dump, while a popular web site called "Roma fa schifo" ("Rome is nasty") details all the ways the city fits that description. Rome has also received a new nickname, "Mafia capital," for the nefarious forces believed with good reason to be operating more or less behind the scenes, affecting city services and operations as well as the general climate. One can only hope that Rome will finally find the leadership it needs to emerge from these conditions, but it is also the case that some problems are systemic, transcending any one administration. Squalor and splendor have always gone hand in hand in this place, and perhaps always will.

In the interest of making Rome a sustainable city, there have been efforts to improve its mass transit, which is at the bottom of the pile compared to other

Chapter Ten

similarly sized European cities. There have also been calls to impose more restrictions on automobile traffic in the center, reserving larger swathes of it for pedestrians only. Such environmental measures are sorely needed, as Rome was recently ranked the world's tenth most congested city. Whether those measures will be implemented is another question.

Car- and bike-share programs are beginning to gain traction, although the constant vandalization of the bikes has become something of a sport, leading some companies to give up on the venture. Like many other Italian cities, Rome already has in place a "limited traffic zone" (*zona traffico limitato*, or ZTL), but the volume of vehicles in the center is still excessive, and the proliferation of illegally parked cars choking crosswalks, sidewalks, and piazzas is a problem of epic proportions. Certainly, pedestrians and bikers, their paths, and their welfare seem to be low on the priority list for civic authorities.

Through all this, the process of urban planning has remained urgent and controversial, as city authorities weigh reverence for Rome's history against the pressing needs of modern life. Throughout its existence, the city has exemplified a tension between preservation and evolution—a complex balancing act familiar to any metropolis seeking to honor its past while remaining relevant in the present and laying a path for the future. This chapter will not attempt to tie a bow on Rome's current and future status, but instead to summarize some of the underlying challenges facing it. Has the city finally begun to falter, as some have claimed—is this the beginning of the end? Or is this most recent chapter just one in Rome's long, complicated, ongoing biography?

Rapid Transit for a Rapidly Changing City

Rome's most recent master plan, as we will see, claims to offer a revolutionary plan of action to propel Rome into its fourth millennium of life. To that end, it is heavily invested in mass transit, its designers gambling that the city's transport system will be up to the task of carrying Romans into the future—and into the suburbs. Until recently, that would have been a losing bet. Since Mussolini's reform of 1930, Roman transport has relied heavily on an extensive bus system, which—while preferable to the surge in automobile traffic and smog that would follow if all those passengers were instead driving cars or scooters—still clogs city streets and suffers from problems of maintenance, crowding, and general dysfunction. In the meantime, Rome has fallen short when it comes to rapid transit, lacking what almost all great, working cities have in common: an extensive underground or subway system that ferries people within and around the city as well as to its periphery and even beyond, without clogging up the surface.

Until just a few years ago, the city's *Metropolitana*, or Metro, consisted of just two lines that intersected at Termini train station, with a scant few stops in the center (fig. 138). The original line was Metro B, which opened in 1955 and was extended in 1980, then again in 2015. It runs from two separate terminuses at the Rebibbia and Monte Sacro zones northeast of the city to Laurentina at the edge of EUR in the southwest. The line was extended by the northern B1 spur in 2012–15, and further expansion projected beyond Rebibbia is pictured on the map as a dotted, faded continuation. The other main line, Metro A, opened in 1980 and was extended for the millennial Jubilee of 2000. It runs between the Battistini station west of the Vatican to the Anagnina station in the southeastern suburbs (near the ubiquitous IKEA, familiar to urban fringes everywhere in the Western world).

Each Metro line has some stops within or along the Aurelian Wall. Other than Termini, Metro A has stations along the northeast of the city, at Piazza della Repubblica (although as of this writing that station has been closed for months due to an escalator collapse resulting in multiple injuries), Piazza Barberini, Piazza di Spagna, and Piazza del Popolo. Metro B penetrates the core of the city more directly. After Termini, it stops at Via Cavour, the Colosseum, the Circus Maximus, and the Piramide station by Porta San Paolo, Rome's old southernmost city gate. Basically, the Metro system in this "X" configuration has worked well—for the small fraction of people who happen to live and work near its stops. The majority who do not must take their chances with the buses, or perhaps the occasional tram.

The long awaited, fully automated, and modern Metro C would seem to hold the promise of major improvement to Rome's rapid transit. It was begun in 1990, and sections of the line lying southeast of Rome finally opened in 2014 and 2015. In 2018, with considerable fanfare, it reached San Giovanni: the first stop at the Aurelian Wall, where the line intersects with Metro A and joins the network as a whole. Within that perimeter, however, the construction of Metro C has advanced at a glacial pace and with much controversy.

Plagued by planning difficulties and cost overruns, its realization has been further complicated by the fact that each meter of tunnel dug seems to yield an archaeological discovery and put a halt to construction. Federico Fellini seemed to predict this scenario in a scene from his 1972 film *Roma*, where the digging of a subway tunnel unearths fabulous ancient frescoes—which proceed to dissolve upon contact with the air. Critics of the plan to bring Metro C into the city center are quick to point out that there is a reason Rome's subway has only minimally entered the Aurelian Wall: the soil beneath the surface is alluvial and full of ruins.

If by some miracle (or travesty) Metro C were completed as planned—and the subway map illustrated here shows its projected continuation as a dotted light-green trail extending to the left of San Giovanni—it would pass right through the center of the city, intersecting with Metro A at the Colosseum stop and running for a time beneath the route of Corso Vittorio: stopping at Piazza Venezia, Chiesa Nuova, and St. Peter's. As the map shows, there are even plans for a Metro D—pictured as a dotted orange line—and rumors of a Metro E . . . but first things first. Metro C is still a work in progress, and the recent map of the official plans for Rome's Metro system pictured here promotes a utopic urban future where people pass fluidly in and out of the city center to the surrounding quarters, where all is efficiency and seamless connectivity, where no messy (or wonderful) urban realities interfere with modernization.

The map suggests not only that the subway will be funded—which seems implausible given that Rome's mass transit administration, ATAC, is over a billion euros in debt and recently declared bankruptcy—but also that it will not encounter (or indulge) any obstacles. This misbegotten plan treats the city as a blank slate while disregarding its very real history, much of which lies below ground. In short, the map is just as idealized as any discussed earlier in this book, its grand, reckless plans verging on the fantastical.

A Master Plan for the Third Millennium: (Un)sustainable Rome

In thinking of the city's future, it is clear that any potential for growth lies beyond the historical core. That is the premise of the latest master plan (or *piano regolatore*; fig. 139), which promises a new vision for Rome's evolution, but which is by no means a certain plan of action. Rome has become many things since 753 BCE, but one of them is *not* a planned city—in fact, the city seems to thwart such efforts. Since the first master plans of 1873 and 1883, more have followed: in 1909, 1931, 1962 (amended in 1967 and 1974), and 2008 (amended 2016). Despite the good intentions behind most of these maps, their execution, left to others, has been piecemeal at best. Each master plan operates a bit like an optimistic statement of best practices—put it out there and hope people heed the guidelines.

Like previous versions, the most recent one must also contend with a fundamental conundrum: it aims to impose order retrospectively on areas that have developed in absence of prior planning. Each master plan has followed on the heels of a period of major expansion, rather than laying the groundwork for one—so each is trying to put the proverbial horse back in the barn. That said, the 2008 master plan represents a major paradigm shift in its basic approach to Rome. No longer is the city conceptualized as what it was for most of its long existence: a single epicenter from which development radiates outward toward the periphery. The whole notion of center and periphery has been replaced by a polycentric model, a notion of Rome as a "city of cities" functioning as an integrated network. In this scheme, Rome's identity as a walled precinct is finally obsolete. The Aurelian Wall remains in place, but the city is not defined by physical boundaries.

The connections between Rome's multiple centers are to be facilitated, of course, by rapid transit—not, one hopes, an increase in automobile traffic (see the purple areas denoting rail and subway zones in fig. 140). Indeed, the possibility of this plan coming to fruition is fully dependent on the ongoing expansion of Rome's Metro, including the purely hypothetical Line D as well as the fragmentary and halting Line C. The extent to which transit is a pillar of the

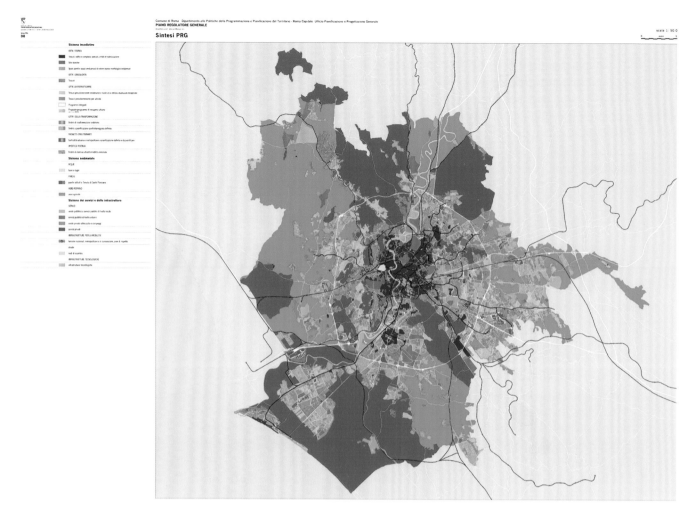

Fig. 139

Above: *Piano regolatore generale* (master plan of Rome), synthesis map, 2008. © Comune di Roma.

Fig. 140

Opposite: *Piano regolatore generale* (master plan of Rome), map of strategic planning areas, 2008. © Comune di Roma.

new master plan is also evident in the premise that new development will only be permitted if it is connected to an existing or projected station, and that new, concentrated zones (or metropolitan centralities) are to be created around the most important stations.

Although its dependence on the appearance of new Metro lines might be ill-advised, the 2008 master plan is, in a sense, a nod to reality: Rome's expansion beyond the Aurelian Wall surged after World War II, and the city continues to gobble up more and more of the surrounding region of Lazio. Peripheral centers have already arisen, many of them squalid and/or isolated, so the theory goes that now it is a matter of connecting them and controlling development in and around them. To that end, the latest master plan also calls for protecting the remaining expanses of green space and parkland that are interspersed with more densely populated zones (see the green areas, denoting archaeological parks and parkland, in fig. 140).

Perhaps most fundamental and praiseworthy, it proposes ecological conservation and recovery of the whole Tiber River valley upon which all these spaces depend (see the aqua-colored zones in fig. 140). With its professed, progressive

Chapter Ten

Comune di Roma

PIANO REGOLATORE GENERALE

adottato con del. C.C. n.33 del 19/20 marzo 2003

Ambiti di programmazione strategica: quadro di unione

D7

Comune di Roma   Dipartimento alle Politiche della Programmazione e Pianificazione del Territorio – Roma Capitale   Ufficio Pianificazione e Progettazione Generale

PIANO REGOLATORE GENERALE
Direttore arch. Daniel Modigliani

scala 1: 20.000

0   metri   2.000

Ambiti di programmazione strategica: quadro di unione

Ambito Tevere

Ambito Mura

Ambito Parco archeologico - monumentale
dei Fori e dell' Appia antica

Ambito Flaminio Fori Eur

Ambito Cintura ferroviaria

emphasis on environmentalism, reduction of automobile traffic, improvements in mass transit, and multiple interlinked clusters, the 2008 master plan holds out the promise of a Rome that will continue to stand the test of time.

All that said, there is plenty of cause for cynicism. Rome is decentralized, critics argue, not because of unplanned development that should be integrated but in most instances because of calculated, mercenary development, namely the creation of huge shopping centers in the middle of nowhere like Romaest, Euroma, and Porta di Roma. By proposing to incorporate such ventures into Rome's fabric, does the latest master plan seek to nurture the city's long-term health or line the pockets of property speculators? Only time will tell if this master plan is another naïve utopia, a more calculated, rapacious venture that throws Rome's future under a (flaming) bus, or a viable path forward.

Whatever the answer, this plan or any other would have to contend with an almost hardwired Roman tendency toward speculative property development—which tends to favor low-density suburban sprawl, inadequately served by public transport—and in which the prime movers and beneficiaries are often connected to the city's governance and planning. That major issue aside, many questions remain: Will competing factions and stakeholders be able to align their interests and bring this or any other plan to fruition? Integrated city planning has never been Rome's forte. Can people even agree on what constitutes sustainability? Or will divisions and disagreements lead to further unchecked, and unsound, growth? What then?

In 2012, the late Renato Nicolini—architect, politician, and onetime mayoral candidate of Rome—wrote an editorial for the leftist newspaper *Il Manifesto* denouncing what he perceived to be the greed and empty promises at the heart of the latest master plan. Rome's governors, he wrote, need to turn back to the city's core identity, which involves something more "immaterial" and "complex" than any physical feature: its culture, which "courses through the feeling of citizenship and of belonging." Rome, he suggested, can remain relevant in our global time and compete with other, larger cities with its "museums and monuments, with its landscape and history, its training, research and creativity, imagination, pleasure of living." However vague those notions, if he is right, then there is reason to take heart. Rome, in the end, is and has always been about much more than brick and mortar, walls, churches, ruins, and relics. There is no way to draw a fine point on the sublime mess, the *bel casino*, that is Rome, but it seems safe to say that one way or another, the Eternal City will live on.

FURTHER READING

Bruscoli, Beatrice. "Terrain Vague: The Tiber River Valley." *Waters of Rome* 10 (2016): 1–31, http://www3.iath.virginia.edu/waters/Journal10Bruscoli_mini.pdf.
Colombo, Andrea. *Marcio su Roma: Criminalità, corruzione e fallimento della politica nella capitale*. Milan: Cairo Publishing, 2016.

Delpirou, Aurélien. "Transport and Urban Planning in Rome: An Unholy Marriage?" *Metropolitics* 4 (2012), https://www.metropolitiques.eu/Transport-and-urban-planning-in.html.

Gemmiti, Roberta, Luca Salvati, and Silvia Ciccarelli. "Global City or Ordinary City? Rome as a Case Study." *International Journal of Latest Trends in Finance & Economic Sciences* 2, no. 2 (2012): 91–98.

Giagni, Tommaso. "La Roma abbandonata dei centri commerciali." *L'Espresso*, June 22, 2018.

Malandrino, Cosima. "Rome's Transportation Crisis: An Overview Ahead of the Referendum." *Urban Media Lab*, October 26, 2018, https://labgov.city/theurbanmedialab/romes-transportation-crisis-an-overview-ahead-of-the-referendum/.

Marcelloni, Maurizio. *Pensare la città contemporanea: Il nuovo piano regolatore di Roma.* Rome: Laterza, 2003.

Nicolini, Renato. "Cambiamo Roma, sono pronto." *Il Manifesto*, June 28, 2012.

Salvati, Luca. "Towards a Polycentric Region? The Socioeconomic Trajectory of Rome, an 'Eternally Mediterranean' City." *Tijdschrift voor Economische en Sociale Geografie* 105, no. 3 (2014): 268–84.

Acknowledgments

Mย first debt of gratitude goes to the city of Rome itself, an amazing and inspiring subject if ever there was one. On a more personal level, this book owes a lot to my son Marco, whose 2017 arrival prompted me to embark on this "fun" project as a way to keep my brain working while caring for an infant. It has been a joy to watch him grow along with the book. He and his older brother Matteo have kept me grounded and provided plenty of welcome distraction along the way.

Profound thanks go to my wonderful editor, Mary Laur, who helped to conceptualize the structure, then provided critical feedback and encouragement throughout the writing process, and to the production team at the University of Chicago Press that helped to bring this book to fruition. The content has made significant improvements thanks to the two anonymous readers who read the manuscript carefully and provided substantive suggestions.

I am grateful to all who provided images, often free of charge or for a nominal fee. Most of these sources were not professional photographers or imaging services but rather bloggers, Wikimedians, video game designers, top-notch scholars/archaeologists, cartographers, collectors, and so on. I am also indebted to the major repositories that generously make digital images of many works from their collection freely available, including the Metropolitan Museum of Art, the Getty Research Institute, and Harvard University Art Museums, as well as the innumerable contributors to web platforms such as archive.org and europeana.eu. These individuals and organizations exemplify the benefits of open access in the digital era.

I dedicate this book to my husband Nick Camerlenghi, my resident Rome expert, best constructive critic, and champion, as well as complete and equal partner in work, life, and love.

Index

Page numbers in italics refer to figures.

Index

Index